SpringerBriefs in Applied Sciences and Technology

For further volumes:
http://www.springer.com/series/8884

Koji Sugioka · Ya Cheng

Femtosecond Laser 3D Micromachining for Microfluidic and Optofluidic Applications

 Springer

Koji Sugioka
Laser Technology Laboratory
RIKEN
Saitama
Japan

Ya Cheng
State Key Laboratory of High Field
 Laser Physics
Shanghai Institute of Optics
 and Fine Mechanics
Chinese Academy of Sciences
Shanghai
People's Republic of China

ISSN 2191-530X   ISSN 2191-5318 (electronic)
ISBN 978-1-4471-5540-9   ISBN 978-1-4471-5541-6 (eBook)
DOI 10.1007/978-1-4471-5541-6
Springer London Heidelberg New York Dordrecht

Library of Congress Control Number: 2013948353

Printed on acid-free paper

Springer is part of Springer Science+Business Media (www.springer.com)

Preface

Femtosecond lasers are becoming very common tools for laser materials processing, both for fundamental investigations and various applications including practical uses. Despite this, materials processing using femtosecond lasers does not have a very long history. It was initiated in 1987 when Srinivasan and co-workers and Küper and co-workers demonstrated clean ablation of polymethyl methacrylate (PMMA) with little formation of a heat-affected zone. Furthermore, Küper and co-workers showed that due to their extremely high peak intensities, femtosecond lasers can perform clean ablation of even transparent materials such as NaCl and polytetrafluoroethylene (PTFE) via multiphoton absorption. These experiments had a great impact on many researchers so that research in this field increased rapidly in the 1990s. In 1996, Hirao and co-workers and Mazur and co-workers demonstrated that the interiors of transparent materials such as glasses can be modified or machined by a tightly focused femtosecond laser beam with a moderate pulse energy. This ability was widely used to fabricate three-dimensional (3D) photonic devices such as optical waveguides, optical couplers and splitters, volume Bragg gratings, Fresnel zone plates in glass chips. In 2001, Misawa and co-workers fabricated 3D microfluidic channels in glass by internal modification using a femtosecond laser followed by wet chemical etching. Since 2003, we have been using this ability of femtosecond lasers to perform 3D micromachining inside glass to fabricate biomicrochips such as microfluidics, microreactors, lab-on-a-chip devices, micro-total analysis systems (μ-TAS), and optofluidic devices. Biomicrochips have revolutionized many fields including biochemical analysis since they can be used to perform biochemical analysis with a high efficiency and accuracy and can reduce reagent consumption, waste production, analysis time, and labor costs. We have demonstrated that femtosecond laser processing has many advantages for biomicrochip fabrication over conventional fabrication techniques such as traditional semiconductor processing and soft lithography since it directly fabricates true 3D microfluidic structures and integrates some functional microcomponents including micro-optical, micromechanic, and microelectronic components. Consequently, biomicrochip fabrication is becoming one of the most important and promising applications of femtosecond laser processing and many researchers are working in this field.

Despite the rapid growth in femtosecond laser processing, only a few books have been published on it to date. These books review the fundamental aspects or discuss the wide range of applications of femtosecond laser processing. We considered that a book that focuses specifically on biomicrochips fabricated by femtosecond lasers would be beneficial for researchers working in this and related fields. This book describes biomicrochip fabrication by 3D fabrication techniques inside glass based on femtosecond lasers. It contains many illustrations and photographs cited from papers published by ourselves and other groups. It comprises 10 chapters, namely the introduction (Chap. 1), current techniques for fabricating microfluidic and optofluidic components (Chap. 2), fundamentals of femtosecond laser processing (Chap. 3), fabrication of microfluidic structures (Chap. 4), fabrication of fluid control microdevices (Chap. 5), fabrication of micro-optical components (Chap. 6), selective metallization (Chap. 7), integration of functional microcomponents Chap. 8), applications of biomicrochips for biological analysis (Chap. 9), and a summary and outlook for this field (Chap. 10). This book thus provides a comprehensive review of the state of the art and future prospects as well as fundamental aspects of this field. We hope that this book will be beneficial for students and young scientists who either are considering working or have just started working in femtosecond laser processing or biomicrochips as well as for researchers and engineers in both academia and industry who are already working in these fields.

Finally yet importantly, we would like to thank Shoji Maruo, Andreas Ostendorf, Malcolm Gower, Yoshiki Nakata, Shigeki Matsuo, Roberto Osellame, Costas Grigoropoulos, and Yves Bellouard for allowing us to use their original figures in this book. We would also like to thank Lingling Qiao for her technical assistance in editing this book.

December 2012

Koji Sugioka
Ya Cheng

Contents

1	**Introduction**	1
	References	4
2	**Current Techniques for Fabricating Microfluidic and Optofluidic Devices**	7
	2.1 Introduction	7
	2.2 Glass Etching	9
	2.3 PDMS-Based Soft Lithography	10
	2.4 Femtosecond-Laser-Induced Two-Photon Polymerization	12
	2.5 Femtosecond Laser Direct Writing in Glass	14
	2.6 Summary	15
	References	16
3	**Fundamentals of Femtosecond Laser Processing**	19
	3.1 Introduction	19
	3.2 Principles and Characteristics of Femtosecond Laser Processing	20
	3.2.1 Nonthermal Processing	20
	3.2.2 Minimized Heat-Affected Zone	21
	3.2.3 Nonlinear Absorption (Photoionization)	21
	3.2.4 Internal Modification of Glass	23
	3.2.5 Enhanced Spatial Resolution	24
	3.3 Interaction of Femtosecond Laser Radiation with Glass	25
	3.4 State-of-the-Art Femtosecond Laser Processing	26
	3.5 Summary	30
	References	30
4	**Fabrication of Microfluidic Structures in Glass**	35
	4.1 Introduction	35
	4.2 Femtosecond-Laser-Assisted Wet Chemical Etching	36
	4.2.1 Fabrication in Photosensitive Glass	36
	4.2.2 Fabrication in Fused Silica	38
	4.3 Liquid-Assisted Femtosecond Laser Three-Dimensional Drilling	41

4.4   Shaping Techniques for Fabricating Microfluidic Channels . . .   44
4.5   Summary   46
References   47

5   **Fabrication of Fluid Control Microdevices**   49
5.1   Introduction   49
5.2   Microvalve Fabrication   50
5.3   Microrotor Fabrication   52
5.4   Summary   54
References   55

6   **Fabrication of Micro-optical Components in Glass**   57
6.1   Introduction   57
6.2   Optical Waveguides   58
6.3   Micro-optical Mirrors   60
6.4   Micro-optical Lenses   64
6.5   Micro-optical Cavities   65
6.6   Micro-optical Attenuators   69
6.7   Summary   71
References   71

7   **Selective Metallization of Glass**   75
7.1   Introduction   75
7.2   Femtosecond-Laser-Assisted Electroless Plating   76
7.3   Femtosecond Laser Modification Combined
        with Electroless Plating   78
        7.3.1   Modification Using Silver Nitrate   79
        7.3.2   Modification by Ablation   81
7.4   Fabrication of Three-Dimensional Metal Microstructures   83
7.5   Summary   86
References   86

8   **Integration of Microcomponents**   89
8.1   Introduction   89
8.2   Integration of Free-Space Micro-optical Components
        and Optical Waveguides   90
8.3   Microfluidic Dye Laser   91
8.4   Mach–Zehnder Optical Modulator   95
8.5   Optofluidic Systems   97
8.6   Summary   101
References   101

9   **Applications of Biochips Fabricated by Femtosecond Lasers**   105
9.1   Introduction   105

9.2   Biosensing Based on Surface-Enhanced Raman
        Scattering Spectroscopy.   106
9.3   Fluid Mixing   107
9.4   Single Cell Detection   108
9.5   Manipulation of Single Cells   110
9.6   Cell Sorting   111
9.7   Concentration Analysis of Liquid Samples   113
9.8   Rapid Screening of Algae Populations   116
9.9   Nanoaquariums for Determining the Functions
        of Microorganisms   117
9.10   Summary.   121
        References   122

10   Summary and Outlook   125
        10.1   Introduction 125
        10.2   Summary. 126
        10.3   Outlook. 126
        References   128

Chapter 1
Introduction

Abstract Femtosecond lasers have opened up new avenues in materials processing due to their unique characteristics of ultrashort pulse widths and extremely high peak intensities. One of the most important features of femtosecond laser processing is that a femtosecond laser beam can induce strong absorption in even transparent materials due to nonlinear multiphoton absorption. Multiphoton absorption enables both surface and internal three-dimensional modification and microfabrication of transparent materials such as glasses. This makes it possible to directly fabricate three-dimensional microfluidic, micromechanic, microelectronic, and micro-optical components in glass. These microcomponents can be easily integrated in a single glass microchip by a simple procedure using a femtosecond laser. Thus, femtosecond laser processing has several advantages over conventional methods such as traditional semiconductor processing or soft lithography for fabricating microfluidic, optofluidic, and lab-on-a-chip devices. Consequently, this topic is currently being intensively studied. This book gives a comprehensive review of the state of the art and future prospects of femtosecond laser processing for fabricating devices such as biomicrochips.

In the last decade, the demand to reduce the volume of samples and reagents used in chemical reactions, biological analysis and medical inspections has increased significantly due to the need to reduce reagent consumption, waste production, analysis time and labor costs. To achieve this, the use of biomicrochip systems such as microfluidic devices, microreactors, lab-on-a-chip devices, micro-total analysis systems (μ-TAS) and optofluidic devices has been proposed [1, 2]. Palm-sized microchips that have the functionalities of an entire laboratory are capable of satisfying this demand. In addition, to achieve highly functional analysis with high efficiency, high accuracy and high sensitivity, increasingly complex microchips should be fabricated in which microfluidic-based systems are integrated with electrical, mechanical and optical microcomponents in a single chip.

Microchip fabrication currently relies on planar microfabrication techniques such as soft lithography or conventional semiconductor processes based on photolithography (see Chap. 2). Earlier techniques for micro and nanofabrication employ nonlithographic direct patterning based on replica molding [3]. Polydimethylsiloxane

K. Sugioka and Y. Cheng, *Femtosecond Laser 3D Micromachining for Microfluidic and Optofluidic Applications*, SpringerBriefs in Applied Sciences and Technology, DOI: 10.1007/978-1-4471-5541-6_1, © The Author(s) 2014

(PDMS) is commonly used as a substrate for fabricating biomicrochips by this technique [4]. Although soft lithography is a well-established, rapid and cost-effective technique and hence is suitable for surface microfabrication, it cannot directly form microfluidic structures inside PDMS. To embed microstructures in microchips, fabricated surface microfluidic structures should be covered with another PDMS plate. In addition, PDMS is chemically incompatible with many organic solvents and compositional inhomogeneities frequently cause optical scattering, which is detrimental for optofluidic applications.

Conventional semiconductor processes, which are also well-established techniques, are employed for fabricating microfluidic devices in Si or glass substrates. They were originally developed for manufacturing Si integrated circuits but can be easily modified for fabricating Si-based biomicrochips [5]. Another advantage of using Si as a substrate is that different components such as microelectrodes, photodetectors and ion-sensitive field effect transistors can be integrated. However, Si is opaque at visible wavelengths, conductive and bio-incompatible, making it unsuitable for some applications. Glass is an excellent alternative material as substrates for biomicrochips due to its high chemical resistance to most acids and alkalis, high transparency over a wide wavelength range ranging from ultraviolet to infrared wavelengths, and a high performance over temperatures from less than −100 °C to over 500 °C. However, it is more expensive than polymers and PDMS. Conventional semiconductor processes are also commonly used for fabricating biomicrochips based on glass. Another widely used process is wet chemical etching in hydrofluoric (HF) acid, which is suitable for fabricating relatively large structures [6]. Both conventional semiconductor processes and wet etching require using photolithography for surface patterning. Microfluidic channels are embedded in a glass microchip by simply sealing the microchip with PDMS. However, since the adhesion is too low for some applications (despite PDMS having a high adhesiveness), an adhesive is sometimes used. However, the adhesive could leak into the microfluidic structures embedded in the microchip, contaminating samples. Furthermore, complex fabrication procedures involving multilayer and multistep processes for stacking and bonding substrates are required to construct complex three-dimensional (3D) microstructures.

Femtosecond laser processing is a promising technique for fabricating biomicrochips since it can modify the interior of glass in a spatially selective manner and hence it can directly form 3D microfluidic structures in substrates [7–9]. The extremely high peak powers generated can induce strong absorption even in transparent materials such as glass due to nonlinear multiphoton absorption [10, 11], as is discussed in Chap. 3. By focusing a femtosecond laser beam with a moderate pulse energy inside glass, multiphoton absorption can be confined to a region near the focal point where the laser intensity exceeds a critical value above which multiphoton absorption occurs efficiently and consequently internal modification and machining of transparent materials can be realized [12, 13]. Femtosecond laser pulses modify the chemical properties of the substrate in the laser-irradiated regions. These regions can then be selectively removed by successive wet etching using acids such HF acid, realizing direct fabrication of 3D microfluidic devices [14–16]. This two-step process (i.e., femtosecond laser direct writing followed by wet etching) can also be used to integrate micromechanical components such as microvalves [17] and micropumps [18] to control fluid flow in microfluidic devices. This fabrication process can be extended to fabricate free-space optics such as micromirrors and micro-optical lenses inside glass [19–21].

Another scheme for fabricating 3D microfluidic structures in glass materials is liquid-assisted femtosecond laser drilling in which a distilled water always touchs with the laser-irradiated region is used to efficiently remove ablated debris [22–25]. Femtosecond laser direct writing can also alter the optical properties (e.g., refractive index and optical transmissivity) of the substrate in addition to the chemical properties. For instance, internal refractive index modification has been successfully applied to fabricate a wide range of micro-optical components inside glass materials, including optical waveguides, optical couplers and splitters, volume Bragg gratings and Fresnel zone plates [12, 26–33]. The unique ability of femtosecond laser direct writing to alter both the chemical and optical properties of glass materials enables various biomicrochips to be fabricated through easy integration of various functionalities. The integration of microfluidic and micro-optical components in a single glass chip was used to fabricate a microfluidic dye laser, which can be used as a light source for photonic biosensing in microchips due to its tunability in the visible range [34]. Integration of optical waveguides and a micromirror made of a hollow microplate inside glass has been used to bend a laser beam at an angle of 90° with a bending loss smaller than 0.3 dB in a small chip [35]. Optical waveguides and micro-optical components can be further integrated with microfluidics for manufacturing highly sensitive optofluidic devices [36]. Moreover, femtosecond laser irradiation in combination with electroless plating can selectively metallize glass surfaces, even the internal walls of microfluidic structures [37, 38]. Such selective metallization is useful for forming microelectrodes to electromagnetically control the motion of micromechanical components and electrochemically analyze samples in microfluidic devices.

Biomicrochips fabricated by femtosecond lasers have been used for biological analysis, such as nanoaquariums for determining the functions of living microorganisms [7, 39, 40] and optofluidic sensors possessing various functionalities for sensing the concentrations of liquid samples [41, 42], detecting and manipulating single cells [43–45] and rapidly screening algae populations [46–48].

Despite the high performance of femtosecond lasers for fabricating bio-microchips and the rapid growth in this field, few books have reviewed the fabrication of microfluidic and optofluidic devices by femtosecond laser processing. This book provides a comprehensive overview of this area and includes reviews on current techniques for fabricating microfluidic and optofluidic devices (Chap. 2), fundamentals and characteristics of femtosecond laser processing (Chap. 3), and fabrication of microfluidic structures (Chap. 4), fluid control microdevices (Chap. 5), micro-optical components (Chap. 6) and microelectronics (Chap. 7) in glass by femtosecond laser processing. Integration of such functional microcomponents in a single glass chip (Chap. 8) and applications of such integrated microchips for biological analysis (Chap. 9) are then introduced. Finally, a summary and outlook for this field are given (Chap. 10).

References

1. Burns MA, Johnson BN, Brahmasandra AN et al (1998) An integrated nanoliter DNA analysis device. Science 282:484–487
2. Dittrich PS, Tachikawa K, Manz A (2006) Micro total analysis systems. Latest advancements and trends. Anal Chem 78:3887–3907
3. Xia Y, Whitesides GM (1998) Soft lithography. Annu Rev Mater Sci 28:153–184
4. McDonald JC, Whitesides GM (2002) Poly (dimethylsiloxane) as a material for fabricating microfluidic devices. Acc Chem Res 35:491–499
5. Burg TP, Mirza AR, Milovic N et al (2006) Vacuum packaged suspended microchannel resonant mass sensor for biomolecular detection. IEEE/ASME J Microelectromech Syst 15:1466–1476
6. Tokeshi M, Minagawa T, Uchiyama K et al (2002) Continuous-flow chemical processing on a microchip by combining microunit operations and a multiphase flow network. Anal Chem 74:1565–1571
7. Sugioka K, Hanada Y, Midorikawa K (2010) Three-dimensional femtosecond laser micromachining of photosensitive glass for biomicrochips. Laser & Photon Rev 4:386–400
8. Sugioka K, Cheng Y (2011) Integrated microchips for biological analysis fabricated by femtosecond laser direct writing. MRS Bull 36:1020–1027
9. Sugioka K, Cheng Y (2012) Femtosecond laser processing for optofluidic fabrication. Lab Chip 12:3576–3589
10. Küper S, Stuke M (1989) Ablation of polytetrafluoroethylene (Teflon) with femtosecond UV excimer laser pulses. Appl Phys Lett 54:4–6
11. Küper S, Stuke M (1989) Ablation of uv-transparent materials with femtosecond uv excimer laser pulses. Microelectron Eng 9:475–480
12. Davis KM, Miura K, Sugimoto N et al (1996) Writing waveguides in glass with a femtosecond laser. Opt Lett 21:1729–1731
13. Glezer EN, Milosavljevic M, Huang L et al (1996) Three-dimensional optical storage inside transparent materials. Opt Lett 21:2023–2025
14. Marcinkevicius A, Juodkazis S, Watanabe M et al (2001) Femtosecond laser-assisted three-dimensional microfabrication in silica. Opt Lett 26:277–279
15. Masuda M, Sugioka K, Cheng Y et al (2003) 3-D microstructuring inside photosensitive glass by femtosecond laser excitation. Appl Phys A 76:857–860
16. Cheng Y, Sugioka K, Midorikawa K (2005) Three-dimensional micromachining of glass using femtosecond laser for lab-on-a-chip device manufacture. Appl Phys A 81:1–10
17. Masuda M, Sugioka K, Cheng Y et al (2004) Direct fabrication of freely movable microplate inside photosensitive glass by femtosecond laser for lab-on-chip application. Appl Phys A 78:1029–1032
18. Matsuo S, Kiyama S, Shichijo Y et al (2008) Laser microfabrication and rotation of ship-in-a-bottle optical rotators. Appl Phys Lett 93:051107
19. Cheng Y, Sugioka K, Midorikawa K et al (2003) Three-dimensional micro-optical components embedded in photosensitive glass by a femtosecond laser. Opt Lett 28:1144–1146
20. Cheng Y, Tsai HL, Sugioka K et al (2005) Fabrication of 3D microoptical lenses in photosensitive glass using femtosecond laser micromachining. Appl Phys A 85:11–14

21. Wang Z, Sugioka K, Midorikawa K (2007) Three-dimensional integration of microoptical components buried inside photosensitive glass by femtosecond laser direct writing. Appl Phys A 89:951–955
22. Li Y, Itoh K, Watanabe W et al (2001) Three-dimensional hole drilling of silica glass from the rear surface with femtosecond laser pulses. Opt Lett 26:1912–1914
23. An R, Li Y, Dou Y et al (2005) Simultaneous multi-microhole drilling of soda-lime glass by water-assisted ablation with femtosecond laser pulses. Opt Express 13:1855–1859
24. Liao Y, Ju Y, Zhang L et al (2010) Three-dimensional microfluidic channel with arbitrary length and configuration fabricated inside glass by femtosecond laser direct writing. Opt Lett 35:3225–3227
25. Liao Y, Song J, Li E et al (2012) Rapid prototyping of three-dimensional microfluidic mixers in glass by femtosecond laser direct writing. Lab Chip 12:746–749
26. Yamada K, Watanabe W, Toma T et al (2001) In situ observation of photoinduced refractive-index changes in filaments formed in glasses by femtosecond laser pulses. Opt Lett 26:19–21
27. Schaffer CB, Brodeur A, Garcia JF et al (2001) Micromachining bulk glass by use of femtosecond laser pulses with nanojoule energy. Opt Lett 26:93–95
28. Bricchi E, Mills JD, Kazamsky PG et al (2002) Birefringent Fresnel zone plates in silica fabricated by femtosecond laser machining. Opt Lett 27:2200–2202
29. Kawamura K, Hirano M, Kamiya T et al (2002) Holographic writing of volume-type microgratings in silica glass by a single chirped laser pulse. Appl Phys Lett 81:1137–1139
30. Watanabe W, Kuroda D, Itoh K et al (2002) Fabrication of Fresnel zone plate embedded in silica glass by femtosecond laser pulses. Opt Express 10:978–983
31. Gorelik M, Will S, Nolte A et al (2003) Transmission electron microscopy studies of femtosecond laser induced modifications in quartz. Appl Phys A 76:309–311
32. Watanabe W, Asano T, Yamada K et al (2003) Wavelength division with three-dimensional couplers fabricated by filamentation of femtosecond laser pulses. Opt Lett 28:2491–2493
33. Sudrie L, Winick KA (2003) Fabrication and characterization of photonic devices directly written in glass using femtosecond laser pulsesJ. Lightwave Technol 21:246–253
34. Cheng Y, Sugioka K, Midorikawa K (2004) Microfluidic laser embedded in glass by three-dimensional femtosecond laser microprocessing. Opt Lett 29:2007–2009
35. Wang Z, Sugioka K, Hanada Y et al (2007) Optical waveguide fabrication and integration with a micro-mirror inside photosensitive glass by femtosecond laser direct writing. Appl Phys A 88:699–704
36. Wang Z, Sugioka K, Midorikawa K (2008) Fabrication of integrated microchip for optical sensing by femtosecond laser direct writing of Foturan glass. Appl Phys A 93:225–229
37. Sugioka K, Hongo T, Takai H et al (2005) Selective metallization of internal walls of hollow structures inside glass using femtosecond laser. Appl Phys Lett 86:171910
38. Xu J, Liao Y, Zeng H et al (2007) Selective metallization on insulator surfaces with femtosecond laser pulses. Opt Express 15:12743–12748
39. Hanada Y, Sugioka K, Kawano H et al (2008) Nano-aquarium for dynamic observation of living cells fabricated by femtosecond laser direct writing of photostructurable glass. Biomed Microdevices 10:403–410
40. Hanad a Y, Sugioka K, S-Ishikawa I et al (2008) 3D microfluidic chips with integrated functional microelements fabricated by a femtosecond laser for studying the gliding mechanism of cyanobacteria. Lab Chip 11:2109–2115
41. Crespi A, Gu Y, Ngamsom B et al (2010) Three-dimensional Mach-Zehnder interferometer in a microfluidic chip for spatially-resolved label-free detection. Lab Chip 10:1167–1173
42. Hanada Y, Sugioka K, Midorikawa K (2012) Highly sensitive optofluidic chips for biochemical liquid assay fabricated by 3D femtosecond laser micromachining followed by polymer coating. Lab Chip 12:3639–3688
43. Kim M, Hwang DJ, Jeon H et al (2009) Single cell detection using a glass-based optofluidic device fabricated by femtosecond laser pulses. Lab Chip 9:311–318
44. Bragheri F, Ferrara L, Bellini N et al (2010) Optofluidic chip for single cell trapping and stretching fabricated by a femtosecond laser. J Biophotonics 3:234–243

45. Bellini N, Vishnubhatla KC, Bragheri F et al (2010) Femtosecond laser fabricated monolithic chip for optical trapping and stretching of single cells. Opt Express 18:4679–4688
46. Schaap A, Bellouard Y, Rohrlack T (2011) Optofluidic lab-on-a-chip for rapid algae population screening. Opt Express 2:658–664
47. Schaap A, Rohrlack T, Bellouard Y (2012) Optical classification of algae species with a glass lab-on-a-chip. Lab Chip 12:1527–1532
48. Schaap A, Rohrlack T, Bellouard Y (2012) Lab on a chip technologies for algae detection: a review. J Biophotonics 5:8–9

Chapter 2
Current Techniques for Fabricating Microfluidic and Optofluidic Devices

Abstract A wide variety of techniques have been developed for fabricating microfluidic and optofluidic components and devices using polymer, glass, and silicon substrates. This chapter gives a brief overview of these techniques, which can be categorized into two classes: parallel processing techniques based on photolithography and serial processing techniques based on direct writing. Some representative examples of these two categories are discussed, including photolithography on glass, soft lithography on poly(dimethylsiloxane) (PDMS), and femtosecond-laser-induced two-photon polymerization. The main advantages and disadvantages of parallel and serial processing are compared. Polymers are currently the most commonly used material for microfluidic and optofluidic applications because fabrication in polymers is easy, rapid, and cost effective. In contrast, glass offers better chemical durability and optical performance. Femtosecond laser direct writing enables microfluidic and integrated optofluidic structures with complex three-dimensional geometries to be directly embedded in glass, eliminating the need to use multistep procedures such as stacking and bonding.

2.1 Introduction

As described in Chap. 1, biomicrochips such as microfluidics, lab-on-a-chip devices and micro-total analysis systems (μ-TAS) have enabled high precision control and manipulation of tiny volumes of liquids, permitting high sensitivity analysis of liquid samples and hence facilitating downsizing of both chemical and biological analysis. Early microfluidic systems, such as the gas chromatography system developed by Terry et al. [1] and the miniaturized capillary electrophoresis-based chemical analysis systems developed by Manz et al. [2], were all fabricated by lithography-based micromachining techniques [3]. These techniques were originally developed about 50 years ago for realizing integrated microelectronic chips. In the late 1980s and early 1990s, lithography-based technologies were well developed, enabling micrometer and even sub-micrometer microchannel

K. Sugioka and Y. Cheng, *Femtosecond Laser 3D Micromachining for Microfluidic and Optofluidic Applications*, SpringerBriefs in Applied Sciences and Technology, DOI: 10.1007/978-1-4471-5541-6_2, © The Author(s) 2014

networks to be formed on the surfaces of silicon, glass, and fused silica. A broad range of techniques have subsequently been developed in response to the demand to enhance the functionalities of microfluidic devices and reduce production costs and efficiencies. These fabrication techniques can be broadly categorized into two categories: parallel processing techniques based on photolithography and serial processing techniques based on direct writing.

The first category includes micromachining of glass and semiconductors using photolithography [4–6], soft lithography [7–10], UV laser ablation with mask patterning [11], hot embossing [12], and micro-injection molding [13]. The second category includes direct writing using proton and electron beams [14, 15], laser direct writing [11, 16–21], and xurography (also known as razor writing) employing a cutting plotter [22]. These techniques have been used to produce microfluidic and optofluidic structures and components on a wide range of sub-strates, including glass, semiconductors, various polymers, and recently even on paper [23]. While parallel processing techniques are generally faster than serial processing techniques for producing large quantities of chips, they often require expensive clean-room facilities (e.g., photolithography systems) and specialized skills. In contrast, serial processing techniques permit rapid prototyping of microfluidic structures in a more cost-effective manner when fabricating low quantities of individual components.

With the exception of a very limited number of three-dimensional (3D) fabrication techniques such as femtosecond-laser-induced two-photon polymerization (2PP) [18, 19] and femtosecond laser direct writing in glass [20, 21], all the above-mentioned techniques fabricate device features only on the surfaces of substrates. Thus, subsequent bonding is often required to isolate open features from the environment. Depending on the substrate material, typical bonding techniques include fusion bonding (glass and glass) [24], anodic bonding (silicon and glass) [25], oxygen plasma bonding (PDMS and glass, PDMS and PDMS) [26], and UV curable adhesive bonding [27]. These additional bonding techniques can cause leakage of liquid samples and clogging of thin channels. These problems become more severe when fabricating 3D structures with complex multilayer configurations, since a failure in one layer will affect the performance of the entire device. The following chapters reveal that bonding can be eliminated by directly forming 3D hollow structures in glass materials using femtosecond laser direct writing.

Although there are a wide range of techniques for fabricating microfluidic devices, few of them have been employed for creating integrated optofluidic systems. This is at least partially due to micro-optical component fabrication imposing more stringent requirements on both the substrate material and the surface roughness than microfluidic fabrication. In the remainder of this chapter, we discuss several fabrication techniques that have played important roles in the development of microfluidic and integrated optofluidic devices, including glass etching, soft lithography, and femtosecond-laser-induced 2PP.

2.2 Glass Etching

Silicon was the first substrate material used for microfluidic applications due to its central role in integrated circuit technology. Well-established silicon-based micromachining techniques were directly employed for creating fundamental microfluidic elements such as microfluidic channels and chambers with micrometer and submicrometer fabrication resolutions. However, as a semiconductor, silicon is optically opaque at visible wavelengths and is incompatible with electrokinetic sample transport, making it unsuitable for many microfluidic applications. These problems can be overcome by replacing silicon with other materials such as glasses, ceramics, and polymers.

In the mid-1990s, silicon-based micromachining techniques were extended to the fabrication of microfluidic systems on glass chips [4, 5]. A typical fabrication process is schematically illustrated in Fig. 2.1. Briefly, a glass substrate is first spin-coated with a positive photoresist. After photoresist deposition and pre-baking, the glass substrate is exposed to ultraviolet (UV) radiation through a photomask. The exposed sample is then immersed in a developer to develop the photoresist after baking. In this step, photoresist in the irradiated areas is selectively removed. Next, the glass substrate is immersed in an etchant solution. Microchannel networks are formed on the surface of the glass substrate after etching. Finally, in a glass bonding step, a cover glass is brought into contact with the fabricated glass substrate to form

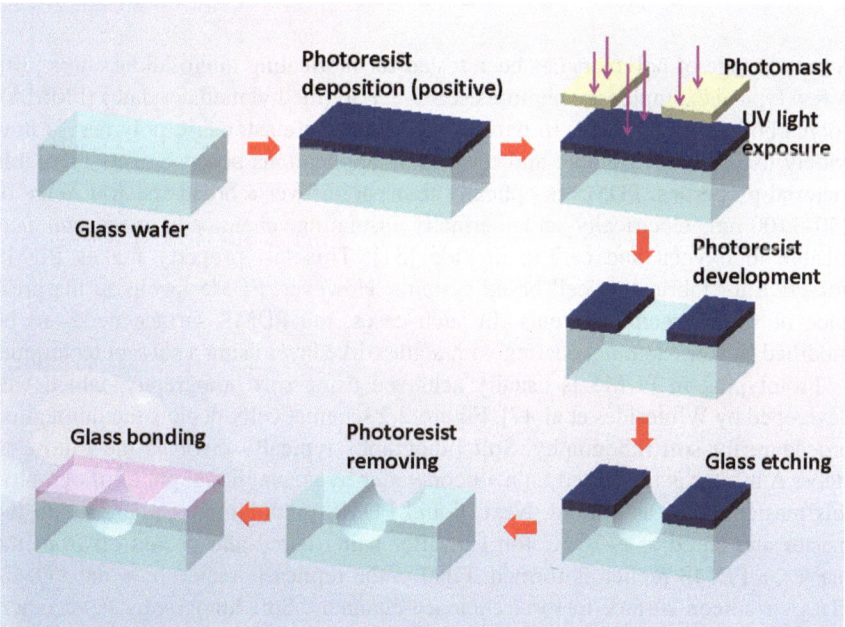

Fig. 2.1 Procedure for fabricating microfluidic devices using glass etching

enclosed channel networks. Glass bonding is commonly achieved by adhesive bonding (also known as glue bonding), anodic bonding, or thermal bonding. Prior to bonding, millimeter-sized connection holes must be made in the cover glass plate; buffer and sample solutions can be injected through these holes into the microchannel networks.

Since wet chemical etching of glass is an isotropic process, it is inherently limited to low-resolution pattern transfer, which produces microchannels with curved sidewalls and low aspect ratios. High-resolution and highly directional pattern transfer can be realized by dry etching techniques such as plasma etching and reactive ion etching, which anisotropically remove substrate material by forming volatile etch products through either ion bombardment or chemical reactions between reactive species combined with physical sputtering by energetic ions. These techniques have been used to fabricate rectangular microfluidic channels in quartz or glass with nanoscale resolutions and aspect ratios greater than 10 [28, 29].

Fabrication of microfluidic devices in glass requires clean-room facilities and specialized skills, which are not easily accessible for many research laboratories. In contrast, polymers are cheaper and their fabrication is faster, more flexible, and more cost-effective. Consequently, they have become the most commonly used substrate material for microfluidic and integrated optofluidic applications.

2.3 PDMS-Based Soft Lithography

A broad range of polymers has been tested for fabricating microfluidic chips [30]. A few typical examples are photoresist SU-8, poly(methyl methacrylate) (PMMA), polycarbonate, and PDMS. In particular, PDMS, an elastomeric polymer, is now widely used for microfluidic and optofluidic applications because of its favorable material properties. PDMS is optically transparent over a broad spectral range of 240–1100 nm, electrically and thermally insulating, chemically inert, and permeable to oxygen and carbon dioxide [31]. This last property makes PDMS attractive for fabricating cell-based systems. However, PDMS swells in the presence of some organic solvents. In such cases, the PDMS surface needs to be modified by, for example, coating with a glass-like layer using a sol–gel technique.

Prototyping in PDMS is usually achieved using soft lithography, which was developed by Whitesides et al. [7]. Figure 2.2 schematically depicts the fabrication procedure for soft lithography. Soft lithography typically involves the following steps. A master is fabricated on a silicon wafer by conventional photolithography; this master serves as a mold. Next, liquid PDMS prepolymer is poured onto the master and cured at ~ 70 °C for 1 h; after being cured and peeled off from the master, a PDMS replica is formed. Finally, the replica is sealed by a flat PDMS, glass, or silicon surface to form enclosed channels. Soft lithography is an inherently 2D fabrication technique because of the photolithography step. However, construction of 3D microfluidic systems can be achieved by stacking multiple thin

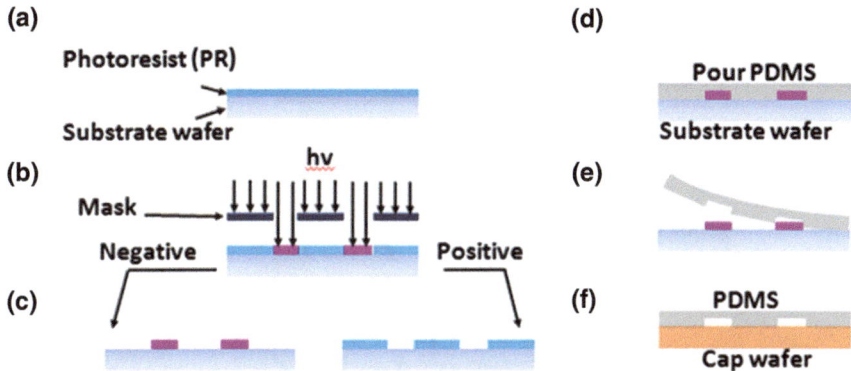

Fig. 2.2 Procedure for fabricating microfluidic structures using soft lithography

2D layers with limited flexibility [32]. The ultimate solution for fabricating 3D microstructures with arbitrary geometries and configurations is to use laser direct writing in transparent substrates based on nonlinear multiphoton absorption [33].

In addition to fabricating microfluidic components, soft lithography can also be used to fabricate micro-optical components in PDMS (although only components with 2D geometries). A micro-optical cylindrical lens has been incorporated in a microfluidic system to improve the excitation efficiency of a fluorescent dye in a microchannel [34]. Figure 2.3a shows the microstructure obtained using SU8 photoresist as a mold; it has a curved sidewall. Figure 2.3b shows a close-up view of a 2D lens fabricated in a PDMS layer. Microfluidic systems integrated with a 2D microlens can also be fabricated in other polymers. For example, a microchip flow cytometer has been fabricated by simultaneously integrating microfluidic channels, waveguides, and micro-cylindrical lens in a negative photoresist SU8 using photolithography [35].

Another strategy for constructing integrated optofluidic systems is to assemble micro-nanofluidic systems, which are prefabricated in PDMS using soft lithography with micro-optical systems fabricated in glass or silicon substrates using conventional photolithography or electron beam lithography. Using this approach, Erickson et al. have demonstrated nanofluidic dynamic tuning of a photonic crystal circuit [36]. This device is schematically depicted in Fig. 2.4a. The bottom layer of the photonic crystal is defined by electron beam lithography and dry etched in a silicon-on-insulator substrate, as shown in Fig. 2.4b. To tune its transmission spectrum, the holes in the central row of the photonic crystal are alternately filled with water and saline, which have different refractive indices. Selective delivery of these liquids to the holes is achieved by covering a nanofluidic channel fabricated in PDMS on top of the photonic crystal, as shown in Fig. 2.4c. The bottom of the nanofluidic channel is punched with an array of uniformly distributed holes, which matches the underlying photonic structure. Fluidic control systems are fabricated at the two ends of the nanofluidic channel; they perform microfluidic functions such as

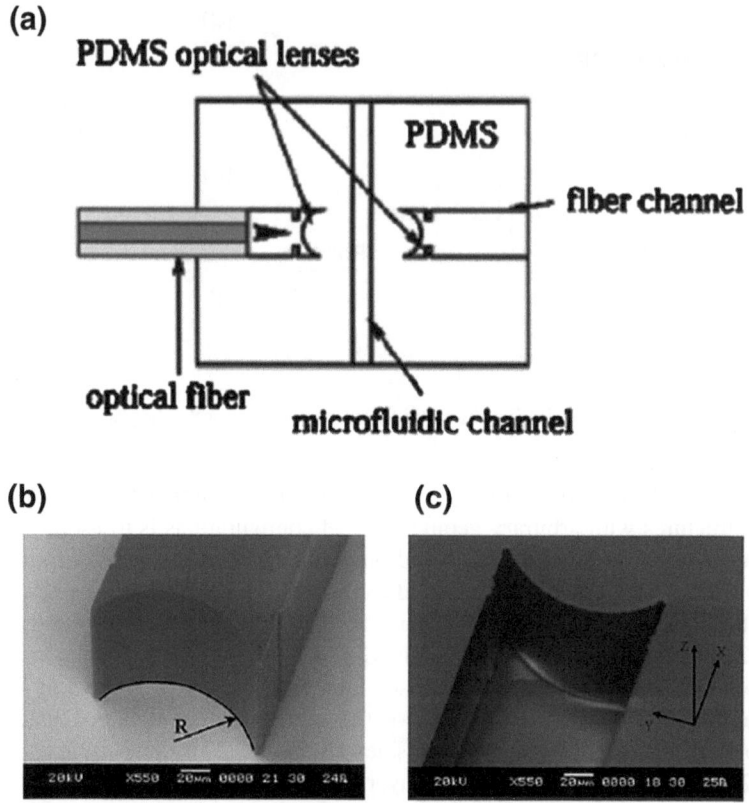

Fig. 2.3 a Schematic diagram of a portable optofluidic device in which intersecting microfluidic channel and optical fibers are integrated. Two microlenses are fabricated near the output and input ends of the fibers to increase the excitation efficiency of a fluorescent dye in a microchannel. **b** Close-up view of SU-8 mold with a curved end, which is a 2D optical lens mold. **c** Close-up view of PDMS layer with a 2D optical lens [34] (Reproduced with permission from RSC; ©2003 by the Royal Society of Chemistry)

mixing and pumping. This device can realize wide tuning of the transmission spectrum on a time scale of seconds.

2.4 Femtosecond-Laser-Induced Two-Photon Polymerization

Femtosecond laser direct writing is an emerging technique for fabricating microfluidic and integrated optofluidic systems that has recently started attracting significant attention. A major advantage of femtosecond laser direct writing is its ability to fabricate 3D structures in transparent materials, including polymers and

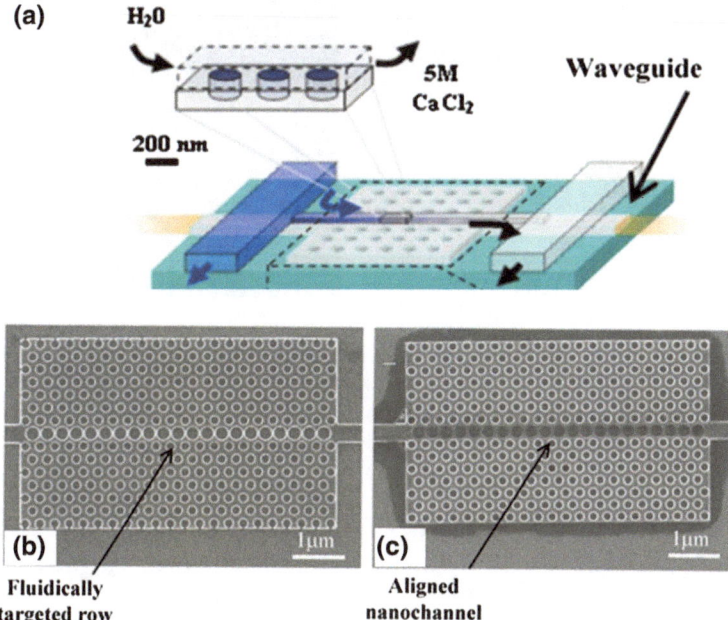

Fig. 2.4 **a** Schematic diagram of a tunable spectral filter realized by nanofluidic targeting of a single row of holes in a planar photonic crystal. Scanning electron microscopy images of **b** photonic crystal prior to assembly and **c** after removing fluidics in nanofluidic channel connected to central row of holes in photonic crystal [36] (Reproduced with permission from OSA ©2006 by the Optical Society of America)

glasses. Since other chapters describe fabrication in glass in detail, this chapter mainly introduces femtosecond laser techniques for fabricating microfluidic and optofluidic structures in polymers. A common approach for forming 3D fluidic and optical structures in polymers is femtosecond-laser-induced 2PP (also known as stereolithography) [37–39], whose concept is illustrated in Fig. 2.5a. 2PP occurs through nonlinear interaction of femtosecond laser pulses with photosensitive resin. It occurs only in the central region of the focal spot where the laser intensity exceeds the 2PP threshold (see Chap. 3). Pre-designed 3D microstructures are written using a focused femtosecond laser beam to convert liquid resin into a solid phase on a point-by-point basis. An important advantage of 2PP is that it can be used for fabricating various passive and active components in microfluidic channels, such as filters [40] and micropumps [41]. Figure 2.5b shows a scanning electron microscopy (SEM) image of a lobed micropump fabricated in a microfluidic channel (width: 5 μm; height: 7 μm) using 2PP [41]. The micropump consists of two counter-rotating rotors, which can be simultaneously optically driven to achieve a pump rate of 0.7 μm/s in the microchannel, as shown in Fig. 2.5c, d. An improved micropump featuring a twin spiral microrotor has

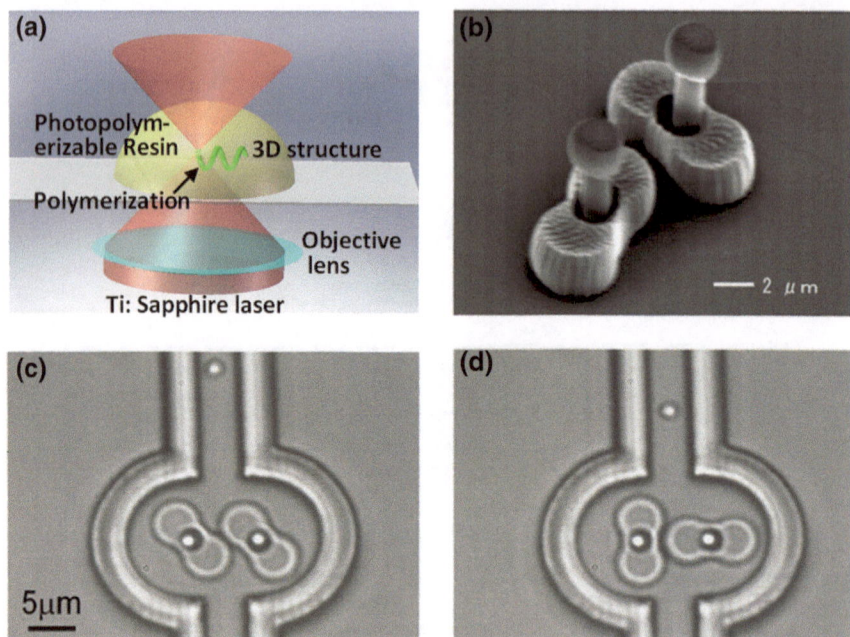

Fig. 2.5 a Illustration of fabrication process for femtosecond-laser-induced 2PP. **b** SEM image of a lobed micropump fabricated using 2PP. **c** and **d** Sequential images taken while optically driving the micropump [41] (courtesy of S. Maruo)

recently been designed and fabricated using 2PP. It has a higher rotation rate than a single spiral microrotor at the same laser power [42].

2PP is an additive fabrication technique. Integrated 3D optofluidic structures can also be formed in polymers by subtractive femtosecond laser fabrication. Wu et al. fabricated an integrated photonic crystal optofluidic sensor on a polymer substrate. They formed a woodpile photonic crystal by creating void channels in PDMS using tightly focused femtosecond laser pulses from a femtosecond laser oscillator [43]. After fabricating the photonic crystal, an open microfluidic channel is produced above the photonic crystal structure by ablating PDMS using a femtosecond laser amplifier. Variations in the refractive index of the liquid in the microfluidic channel can be measured with a sensitivity of $\sim 6 \times 10^{-3}$ by examining the shift in the band gap position of the underlying photonic crystal.

2.5 Femtosecond Laser Direct Writing in Glass

PDMS-based soft lithography remains the most widely used technique for fabricating microfluidic and integrated optofluidic devices due to its low cost, biocompatibility, high mechanical durability, high fabrication resolution, simplicity, and convenience. However, since it is essentially a photolithographic technique, it is inherently planar and hence cannot be used to fabricate 3D microfluidic structures with complex geometries. In addition, as mentioned in Chap. 1, polymers generally have inferior chemical durability and optical performance to most glasses.

Femtosecond laser processing has recently been used for rapid prototyping of complex 3D microfluidic and optofluidic components and devices in glass substrates with extremely high flexibility [42]. Similar to 2PP (see Fig. 2.5a), femtosecond lasers can selectively modify the interior of glass by multiphoton absorption, which occurs only near the focal spot where the laser intensity exceeds a certain threshold. Femtosecond laser irradiation can alter both the optical (e.g., refractive index) and chemical properties (e.g., chemical etch rate) of glass (see Chaps. 4–6 for more details). Thus, waveguides can easily be written by scanning a tightly focused femtosecond beam in glass, while microfluidic channels embedded in glass can be directly formed by femtosecond laser irradiation and subsequent selective wet chemical etching (this technique is frequently referred to as femtosecond-laser-assisted wet chemical etching). Femtosecond-laser-assisted wet chemical etching can also be employed for fabricating free-space optics such as optical mirrors and lenses by creating planar or curved air–glass interfaces in glass. These unique abilities make femtosecond laser direct writing particularly attractive for microfluidic and optofluidic applications.

2.6 Summary

Since the first demonstration of fabrication of microfluidic systems on silicon chips by photolithography and etching, a wide variety of fabrication techniques have been developed for enhancing chip functionalities and reducing production costs and efficiencies. Using these techniques, microfluidic systems have been constructed on a wide range of substrate materials, including polymers, glasses, semiconductors, and even paper. These techniques can generally be categorized into two classes: parallel processing techniques based on photolithography and serial processing techniques based on direct writing. PDMS-based soft lithography is currently the most widely used technique for constructing microfluidic and optofluidic systems. However, it is an inherently 2D technique due to the photolithography step. Consequently, it has limited ability and flexibility for fabricating 3D microstructures with arbitrary geometries.

Femtosecond laser processing has recently been used for rapid prototyping of complex 3D microfluidic and optofluidic components and devices. Femtosecond-laser-induced two-photon polymerization has been used to fabricate various passive and active components (such as filters and micropumps) in microfluidic channels. Compared to polymers, glass offers superior chemical inertness and optical performance. Selective modification of the optical and chemical properties of the interior of glass has been achieved using femtosecond laser irradiation, enabling simultaneous formation of optical waveguides and buried microfluidic channels with arbitrary 3D configurations. The remainder of this book extensively overviews this young yet vibrant research field.

References

1. Terry SC, Jerman JH, Angell JB (1979) A gas chromatographic air analyzer fabricated on a silicon wafer. IEEE Trans Electron Devices ED-26:1880–1886
2. Manz A, Graber N, Widmer HM (1990) Miniaturized total chemical analysis systems: a novel concept for chemical sensing. Sensor Actuat B1:244–248
3. Harrison DJ, Fluri K, Seiler K et al (1993) Micromachining a miniaturized capillary electrophoresis-based chemical analysis system on a chip. Science 261:895–897
4. Grétillat MA, Paoletti F, Thiébaud P et al (1997) A new fabrication method for borosilicate glass capillary tubes with lateral inlets and outlets. Sensor Actuat A 60:219–222
5. Dodge A, Fluri K, Verpoorte E et al (2001) Electrokinetically driven microfluidic chips with surface modified chambers for heterogeneous immunoassays. Anal Chem 73:3400–3409
6. Verpoorte E, Rooij NFD (2003) Microfluidics meets MEMS. Proc IEEE 91:930–950
7. Whitesides GM, Ostuni E, Takayama S et al (2001) Soft lithography in biology and biochemistry. Annu Rev Biomed Eng 3:335–373
8. Zhao XM, Xia YN, Whitesides GM (1997) Soft lithographic methods for nano-fabrication. J Mater Chem 7:1069–1074
9. Xia YN, Whitesides GM (1998) Soft lithography. Annu Rev Mater Sci 28:153–184
10. Unger MA, Chou HP, Thorsen T et al (2000) Monolithic microfabricated valves and pumps by multilayer soft lithography. Science 288:113–116
11. Kim J, Xu XF (2003) Excimer laser fabrication of polymer microfluidic devices. J Laser Appl 15:255–260
12. Becker H, Heim U (2000) Hot embossing as a method for the fabrication of polymer high aspect. Sensor Actuat A 83:130–135
13. Choi JW, Kim S, Trichur R et al (2001) A plastic micro injection molding technique using replaceable mold-disks for disposable microfluidic systems and biochips. In: Proceedings of the 5th international conference on micro total analysis systems (μTAS), pp 411–412
14. Kan JA, Bettiol AA, Watt F (2003) Three-dimensional nanolithography using proton beam writing. Appl Phys Lett 83:1629–1631
15. Mali P, Sarkar A, Lal R (2006) Facile fabrication of microfluidic systems using electron beam lithography. Lab Chip 6:310–315
16. Marcinkevicius A, Juodkazis S, Watanabe M et al (2001) Femtosecond laser-assisted three-dimensional microfabrication in silica. Opt Lett 26:277–279
17. Masuda M, Sugioka K, Cheng Y et al (2003) 3-D microstructuring inside photosensitive glass by femtosecond laser excitation. Appl Phys A 76:857–860
18. Bellouard Y, Said A, Dugan M et al (2004) Fabrication of high-aspect ratio, micro-fluidic channels and tunnels using femtosecond laser pulses and chemical etching. Opt Express 12:2120–2129
19. Osellame R, Hoekstra HJWM, Cerullo1 G et al (2011) Femtosecond laser microstructuring: an enabling tool for optofluidic lab-on-chips. Laser Photonics Rev 5:442–463
20. Schaap A, Rohrlack T, Bellouard Y (2012) Optical classification of algae species with a glass. Lab Chip 12:1527–1532
21. Sugioka K, Cheng Y (2012) Femtosecond laser processing for optofluidic fabrication. Lab Chip 12:3576–3589
22. Bartholomeusz DA, Boutte RW, Andrade JD (2005) Xurography: rapid prototyping of microstructures using a cutting plotter. J Microelectromech Syst 14:1364–1374

23. Li X, Ballerini DR, Shen W (2012) A perspective on paper-based microfluidics: current status and future trends. Biomicrofluidics 6(13):011301

24. Delft KM, Eijkel JCT, Mijatovic D et al (2007) Micromachined Fabry–Pérot interferometer with embedded nanochannels for nanoscale fluid dynamics. Nano Lett 7:345–350

25. Durand NFY, Renaud P (2009) Label-free determination of protein–surface interaction kinetics by ionic conductance inside a nanochannel. Lab Chip 9:319–324

26. Eddings MA, Johnson MA, Gale BK (2008) Determining the optimal PDMS–PDMS bonding technique for microfluidic devices. J Micromech Microeng 18(4):067001

27. Huang Z, Sanders JC, Dunsmor C et al (2001) A method for UV-bonding in the fabrication of glass electrophoretic microchips. Electrophoresis 22:3924–3929

28. He B, Tait N, Regnier FE et al (1998) Fabrication of nanocolumns for liquid chromatography. Anal Chem 70:3790–3797

29. Li X, Abe T, Esashi M et al (2001) Deep reactive ion etching of Pyrex glass using SF plasma. Sensor Actuat A 87:139–145

30. Becker H, Gärtner C (2008) Polymer microfabrication technologies for microfluidic systems. Anal Bioanal Chem 390:89–111

31. McDonald JC, Whitesides GM (2002) Poly (dimethylsiloxane) as a material for fabricating microfluidic devices. Acc Chem Res 35:491–499

32. Anderson JR, Chiu DT, Jackman RJ et al (2000) Fabrication of Topologically Complex Three-Dimensional Microfluidic Systems in PDMS by Rapid Prototyping. Anal Chem 72:3158–3164

33. Liao Y, Song J, Li E et al (2012) Rapid prototyping of three-dimensional microfluidic mixers in glass by femtosecond laser direct writing. Lab Chip 12:746–749

34. Camou S, Fujita H, Fujii T (2003) PDMS 2D optical lens integrated with microfluidic channels: principle and characterization. Lab Chip 3:40–45

35. Wang Z, El-Ali J, Engelund M et al (2004) Measurements of scattered light on a microchip flow cytometer with integrated polymer based optical elements. Lab Chip 4:372–377

36. Erickson D, Rockwood T, Emery T et al (2006) Nanofluidic tuning of photonic crystal circuits. Opt Lett 31:59–61

37. Maruo S, Nakamura O, Kawata S (1997) Three-dimensional microfabrication with two-photon-absorbed photopolymerization. Opt Lett 22:132–134

38. Watanabe M, Sun HB, Juodkazis S et al (1998) Three-Dimensional Optical Data Storage in Vitreous Silica. Jpn J Appl Phys Part 2(37):L1527–L1530

39. Kawata S, Sun HB, Tanaka T et al (2001) Finer features for functional micro-devices. Nature 412:697–698

40. Wang J, He Y, Xia H et al (2010) Embellishment of microfluidic devices via femtosecond laser micronanofabrication for chip functionalization. Lab Chip 10:1993–1996

41. Maruo S, Inoue H (2006) Optically driven micropump produced by three-dimensional two-photon microfabrication. Appl Phys Lett 89(3):144101

42. Maruo S, Takaura A, Saito Y (2009) Optically driven micropump with a twin spiral microrotor. Opt Express 17:18525–18532

43. Wu J, Day D, Gu M (2008)A microfluidic refractive index sensor based on an integrated three-dimensional photonic crystal. Appl Phys Lett 92(3):071108

Chapter 3
Fundamentals of Femtosecond Laser Processing

Abstract Femtosecond lasers have excellent characteristics for materials processing due to their ultrashort pulse widths and extremely high peak powers. When a femtosecond laser beam with a moderate pulse energy is focused into glass, multiphoton absorption or tunneling ionization is confined to a region near the focal point inside the glass. Femtosecond lasers can thus perform internal modification of glass. Internal modification is widely used to fabricate microfluidic structures and micro-optical components, which can be used to produce biomicrochips for biochemical analysis. This chapter reviews the fundamentals and characteristics of femtosecond laser processing. It also introduces state-of-the-art femtosecond laser processing.

3.1 Introduction

The rapid development of femtosecond lasers over the past few decades has opened up new doors for materials processing since such lasers have many advantages over conventional pulsed lasers (i.e., nanosecond lasers). Srinivasan et al. [1] and Küper and Stuke [2] pioneered materials processing using femtosecond lasers in 1987. They demonstrated clean ablation of polymethylmethacrylate (PMMA) with little formation of a heat-affected zone (HAZ) using femtosecond UV excimer lasers. They found that femtosecond lasers have significantly lower ablation thresholds than nanosecond lasers. Furthermore, femtosecond lasers can cleanly ablate even transparent materials such as NaCl and polytetrafluoroethylene (PTFE) by multiphoton absorption because they generate extremely high peak intensities [3, 4]. These experiments had a great impact and consequently research in this field expanded rapidly in the 1990s. Femtosecond lasers are currently becoming very common tools for laser materials processing both in fundamental investigations and for various applications. Such rapid evolution is due to the unique characteristics of femtosecond laser processing as well as the development of high-performance laser systems.

K. Sugioka and Y. Cheng, *Femtosecond Laser 3D Micromachining for Microfluidic and Optofluidic Applications*, SpringerBriefs in Applied Sciences and Technology, DOI: 10.1007/978-1-4471-5541-6_3, © The Author(s) 2014

One important characteristic of femtosecond laser processing is that it reduces heat diffusion to surrounding regions of the processed area [5], resulting in high-quality microfabrication of soft materials such as biological tissues [6] and hard or brittle materials such as semiconductors and insulators [7] without HAZ formation. This suppression of heat diffusion to the surroundings also improves the spatial resolution of nanoscale processing [8]. Additionally, femtosecond laser irradiation at intensities near the ablation threshold produces nanoripple structures on various materials with periodicities much shorter than the irradiation wavelength [9–12]. Another important aspect of femtosecond laser processing is that, as mentioned above, nonlinear absorption (i.e., multiphoton absorption) can induce strong absorption even in materials that are transparent to the laser wavelength [3, 4]. Multiphoton absorption permits not only surface modification but also three-dimensional (3D) internal microfabrication of transparent materials such as glass and polymers [13–16]. The ability of femtosecond lasers to directly form micro-structures inside glass makes them suitable for fabricating microfluidic and optofluidic devices. Additionally, the spatial resolution of multiphoton absorption exceeds the diffraction limit due to the nonlinearity of the process [17].

To understand the factors that make femtosecond laser promising for fabri-cating microfluidic and optofluidic devices, this chapter reviews the fundamentals and characteristics of femtosecond laser processing. It also introduces state-of-the-art femtosecond laser processing.

3.2 Principles and Characteristics of Femtosecond Laser Processing

3.2.1 Nonthermal Processing

In contrast to processing by nanosecond and longer pulses for which thermal processes dominate, femtosecond laser pulses enable nonthermal processing, allowing high-precision material processing to be realized. This characteristic is attributed to rapid energy deposition in the material. It takes a few hundred femtoseconds to a few picoseconds for the electron distribution to reach thermal equilibrium after femtosecond laser irradiation [18, 19]. In contrast, the time to transfer energy from the electron subsystem to the lattice, which induces ther-malization, is of the order of 1–100 ps (depending on the electron–phonon cou-pling strength of the material), which is much longer [20, 21]. Thus, femtosecond lasers can efficiently cause electron heating and generate a hot electron gas, which is far from equilibrium with the lattice. Consequently, only a very small fraction of the laser pulse energy is converted into heat, resulting in nonthermal processing that enables high-quality microfabrication to be performed.

3.2.2 Minimized Heat-Affected Zone

Even though femtosecond laser pulses mainly induce nonthermal processes (see Sect. 3.2.1), they may still generate heat. However, femtosecond laser pulses suppress the formation of a HAZ due to their extremely short pulse widths that are several tens to several hundreds of femtoseconds. This permits high-quality microfabrication, even for metals with high thermal conductivities.

When the laser pulse width is shorter than the electron–phonon coupling time in laser–matter interactions, thermal diffusion to the interior of the material can be almost ignored. Most metals have an electron–phonon coupling time of the order of picoseconds [22], which is sufficiently long compared with the pulse width of femtosecond lasers. In this regime, when the material is heated to near the melting point T_{im} by femtosecond laser irradiation, the thermal diffusion length l_d is given by

$$l_d = \left[\frac{128}{\pi}\right]^{1/8} \left[\frac{DC_i}{T_{im}\gamma^2 C'_e}\right]^{1/4} \tag{3.1}$$

where D is the thermal conductivity, C_i is the lattice heat capacity, C'_e is given by $C'_e = C_e/T_e$ (where C_e is the electron heat capacity and T_e is the electron temperature), and γ is the electron–phonon coupling constant [23]. For example, when copper is heated to its melting point of $T_{im} = 1356$ K by a femtosecond laser, l_d is calculated to be 329 nm [24].

On the other hand, when the laser pulse width τ is much longer than the electron–phonon coupling time, l_d can be approximately estimated using

$$l_d = \sqrt{\kappa\tau} \tag{3.2}$$

Here, κ is the thermal diffusivity. For copper, l_d is estimated to be 1.5 μm for $\tau = 10$ ns. Thus, femtosecond laser processing can clearly reduce the thermal diffusion length, reducing HAZ formation in the processed region.

3.2.3 Nonlinear Absorption (Photoionization)

Another important feature of femtosecond laser processing is that electron excitation (ionization) can be induced even in materials that are transparent to the femtosecond laser beam, through nonlinear processes such as multiphoton absorption (ionization) and/or tunneling ionization.

Figure 3.1 depicts single and multiphoton absorption based on electron excitation. Conventional absorption is linear single-photon absorption. When light whose photon energy exceeds the band gap of a material is incident on the material, it is absorbed and a single photon excites an electron from the valence band to the conduction band. On the other hand, light whose photon energy is smaller than the band gap cannot excite electrons, so that no absorption occurs in

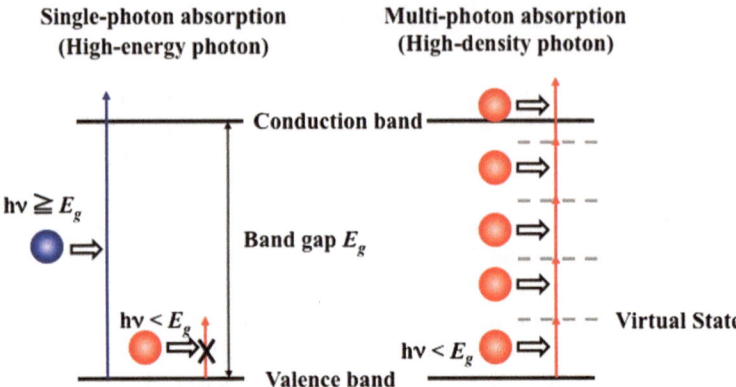

Fig. 3.1 Electron excitation in materials by single and multiphoton absorption

the stationary state. However, when an extremely high density of photons is incident on the material, an electron can be excited by multiple photons, even when the band gap exceeds the photon energy. This process is known as multiphoton absorption. The extremely high density of photons required to induce multiphoton absorption can be easily obtained by femtosecond lasers due to their ultrashort pulse widths.

At high laser intensities and low frequencies, electron excitation is induced by tunneling ionization rather than multiphoton absorption. In tunneling ionization, the potential barrier formed by the valence and conduction band structures is first drastically deformed by the intense electric field of a femtosecond laser and then the barrier length is reduced. When this occurs, an electron can tunnel through the barrier and eventually easily escape from the molecule to be excited from the valence band to the conduction band (see Fig. 3.2).

The probabilities of multiphoton absorption and tunneling ionization in femtosecond laser interaction with transparent materials can be determined by the Keldysh parameter, γ [25]

$$\gamma = \frac{\omega}{e} \sqrt{\frac{m_e c n \varepsilon_0 E_g}{I}} \qquad (3.3)$$

where ω is the laser frequency, I is the laser intensity, m_e is the electron effective mass, e is the fundamental electron charge, c is the speed of light, n is the linear refractive index, ε_0 is the permittivity of free space, and E_g is the band gap of the material. When γ is much greater (smaller) than 1, multiphoton absorption (tunneling ionization) is dominant. For $\gamma \approx 1$, photoionization is induced by a combination of both processes. Thus, femtosecond lasers can induce strong absorption (electron excitation) even in transparent materials, thereby allowing high-quality microprocessing of glass materials.

Fig. 3.2 Tunneling
ionization induced by
femtosecond laser irradiation

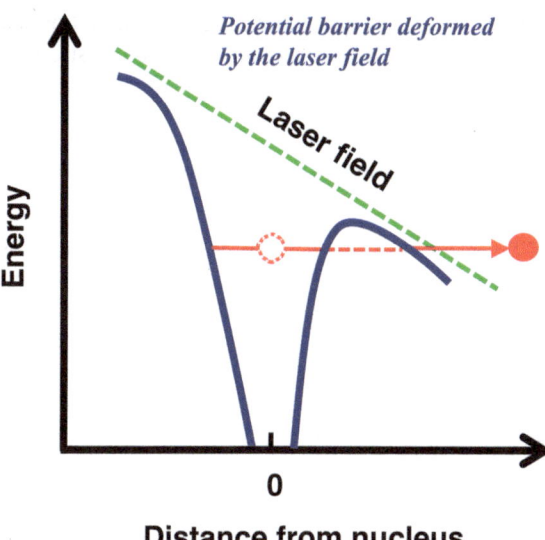

3.2.4 Internal Modification of Glass

Multiphoton absorption and tunneling ionization are nonlinear processes and can be induced only at intensities above a critical value that depends on both the material and the pulse width. When a femtosecond laser beam is focused inside a transparent material with an adequate pulse energy (see Fig. 3.3), absorption can be confined to a region near the focal point inside the material. Thus, internal modification and structure fabrication can be performed in transparent materials. This is possible only using ultrashort pulse lasers such as femtosecond lasers.

Fig. 3.3 Schematic diagram
depicting internal
modification of a transparent
material by multiphoton
absorption or tunneling
ionization induced by
femtosecond laser irradiation

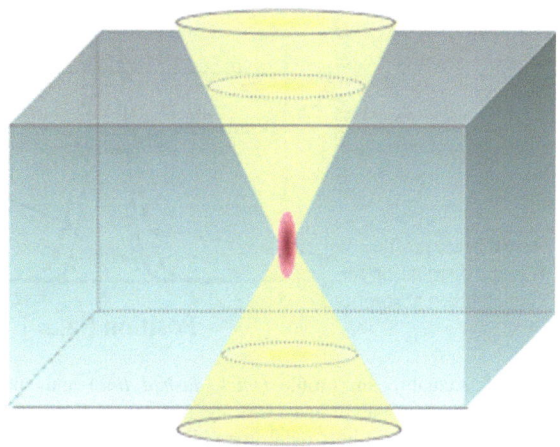

Internal modification can be used to write 3D optical waveguides and to fabricate micro-optical components and microfluidic structures inside glass.

3.2.5 Enhanced Spatial Resolution

Femtosecond lasers can suppress heat diffusion to the surrounding of the processed area, as discussed in Sect. 3.2.2, which is advantageous since it gives higher spatial resolutions. When a 10-ns laser pulse irradiates Cu with a spot size equal to the laser wavelength (typically several hundred nanometers to 1 μm), the processed region becomes larger than the spot size due to the thermal diffusion length of 1.5 μm. In contrast, since there is almost negligible thermal diffusion in femtosecond laser irradiation, the processed region is unlikely to extend beyond the spot size.

Nonlinear multiphoton absorption can further improve the spatial resolution. Ideally, the spatial intensity of a femtosecond laser beam will have a Gaussian profile, as shown by the thick dashed line in Fig. 3.4. For single photon absorption, the spatial distribution of the laser energy absorbed by the material corresponds to this beam profile. However, for multiphoton absorption, the absorbed energy distribution becomes narrower with increasing order (n) of multiphoton absorption, since the effective absorption coefficient for n-photon absorption is proportional to the nth power of the laser intensity. Therefore, the effective beam size ω for n-photon absorption is expressed by

$$\omega = \omega_0/\sqrt{n} \tag{3.4}$$

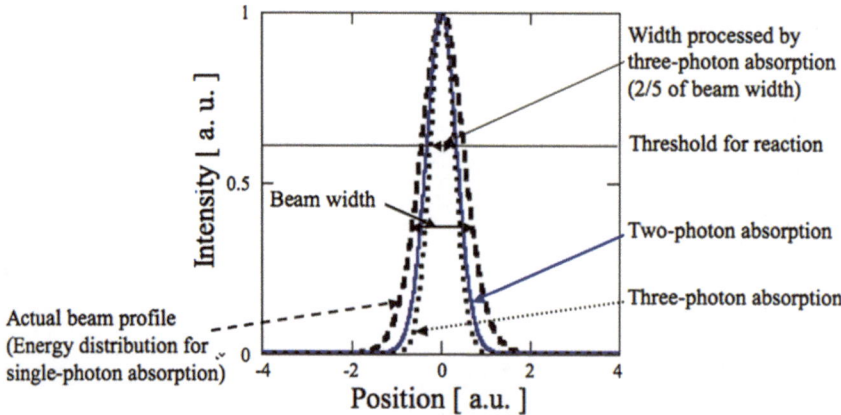

Fig. 3.4 Actual beam profile (*thick dashed line*) and spatial distributions of laser energy absorbed by transparent materials by two (*solid line*) and three (*thin dashed line*) photon absorption. The solid horizontal line indicates the reaction threshold

where ω_0 is the actual spot size of the focused laser beam. Figure 3.4 shows the spatial distributions of the laser energy absorbed by transparent materials in two (solid line) and three (thin dashed line) photon absorption. From Eq. (3.4), the spatial resolution is expected to be much smaller than the wavelength for multi-photon absorption. In addition, when there is a laser intensity threshold above which a reaction occurs after absorption, the fabrication resolution can be further improved by adjusting the laser intensity. For example, if the threshold intensity for the reaction corresponds to the solid straight line in Fig. 3.4, the fabrication width can be reduced to 2/5th that of ω_0. Thus, nonlinear multiphoton absorption can overcome the diffraction limit of the laser wavelength and achieve a subdiffraction resolution.

3.3 Interaction of Femtosecond Laser Radiation with Glass

Femtosecond laser irradiation induces electron excitation and relaxation processes in glass, as depicted in Fig. 3.5 [26]. Electrons are initially excited from the valence band to the conduction band by multiphoton absorption or by tunneling ionization, as described in Sect. 3.2.3. For relatively low laser intensities of the incident femtosecond laser beam, the generated free electrons contribute to photochemical reactions (e.g., photoreduction of ions doped in glass resulting in

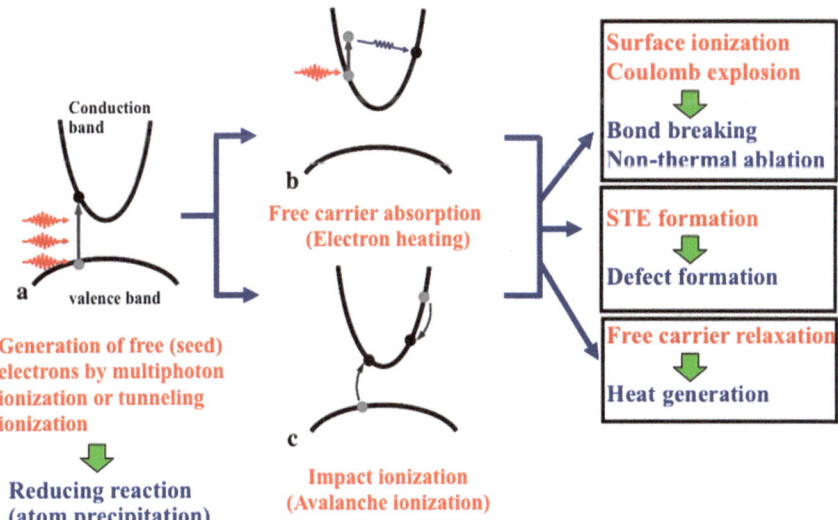

Fig. 3.5 Electron excitation and relaxation processes in glass induced by femtosecond laser irradiation. Only multiphoton absorption is shown for the initial excitation of free electrons

precipitation of atoms). At higher intensities, excited electrons can sequentially absorb several photons in the same laser pulse so that they are excited to higher energy states for which free carrier absorption is efficient. This sequential excitation is termed electron heating and it results in non-thermal bond breaking. At sufficiently high laser intensities, excited electrons are accelerated by the intense electric field of the ultrafast laser beam and collide with surrounding atoms, generating secondary electrons (avalanche ionization), which cause a Coulomb explosion and eventually non-thermal ablation. Some of the generated free electrons relax to localize the energy stored in electron–hole pairs, which form self-trapped excitons (STEs). This relaxation often commences within 1 ps after laser irradiation. Some STEs relax to form permanent defects within a few hundred picoseconds. Glass heating occurs a few tens of picoseconds after laser irradiation due to free electron relaxation and the irradiated area returns to room temperature after several tens of microseconds, causing modification or damage.

3.4 State-of-the-Art Femtosecond Laser Processing

As discussed in Sects. 3.2.1 and 3.2.2, the non-thermal process and reduced HAZ formation of femtosecond laser irradiation permits high-quality, high-precision microfabrication using various materials including metals with high thermal conductivities, soft materials such as biological tissues, and hard or brittle materials such as semiconductors and insulators. Figure 3.6a, b show scanning electron microscopy (SEM) images of holes drilled in 100-μm-thick steel foils by ablation using laser pulses with widths of 200 fs and 3.3 ns, respectively [27]. Femtosecond lasers can also perform high-quality micromachining of glass by multiphoton

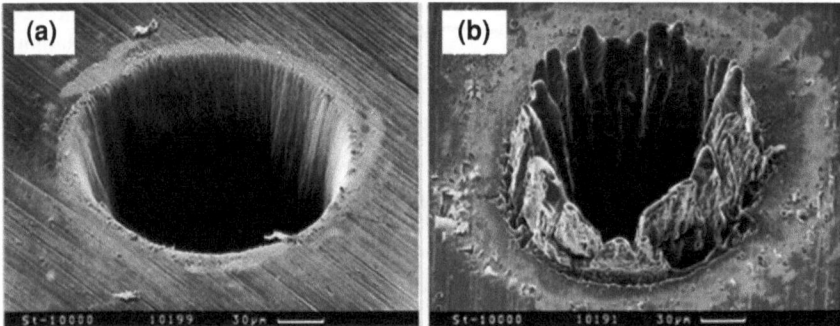

Fig. 3.6 Holes drilled in 100-μm-thick steel foils by ablation using laser pulses with the following parameters: **a** pulse width: 200 fs, pulse energy: 120 μJ, fluence: 0.5 J/cm^2, wavelength: 780 nm; and **b** pulse width: 3.3 ns, pulse energy: 1 mJ, fluence: 4.2 J/cm^2, wavelength: 780 nm. The scale bars represent 30 μm [27] (Reproduced with permission from Springer. ©1996 by Springer)

Fig. 3.7 SEM images of **a** surface micromachining and **b** cutting of glass materials by femtosecond laser ablation (Courtesy of M. Gower)

absorption or tunneling ionization. Figure 3.7a, b respectively show SEM images of surface micromachining and cutting of glass by femtosecond laser ablation. Both images reveal that clean ablation with sharp edges and without cracks was achieved. Such surface micromachining and patterning by laser ablation is attracting growing interest due to their potential use in fabricating devices such as photovoltaic solar panels, flat panel displays, and light emitting devices.

The excellent characteristics of femtosecond laser processing of minimizing HAZ formation enables nanoablation to be performed with subwavelength resolution or smaller [28, 29]. Furthermore, novel nanoscale structures (e.g., nanobumps, nanobelts, and nanomeshes) have been formed on Au thin films deposited on silica by multibeam interference of light from a femtosecond laser [8, 30–35]. In this scheme, it is essential to suppress heat diffusion to surrounding areas from the processed region. Irradiation of a 10-nm-thick gold thin film on a sapphire (0001) substrate by a single pulse from a four-beam-interfering femtosecond laser was found to generate unique nanostructures like a water drop (Fig. 3.8a, b) [34]. This interference pattern had a period of 1.3 μm. The structure was approximately 800 nm high and the bead and neck radii were 75 and 18 nm, respectively. Using a wider-period interference pattern (3.9 μm) and a silicon (100) substrate, which has a higher thermal conductivity than sapphire and is opaque to the laser wavelength, resulted in the formation of nanocrowns (Fig. 3.8c, d) [35]. On the other hand, for a period of 1.7 μm and a sapphire (0001) substrate, nanowhiskers with a sharp top and a curvature radius smaller than 15 nm formed on the bumps (Fig. 3.8e, f) [35]. The spiky structures shown in Fig. 3.8 can generate field enhancements and are promising for use in nanotechnology applications. Despite femtosecond laser processing being generally regarded as a nonthermal technique in which thermal effects are minimized, thermal processes (e.g., melting, flow, bending, and inflation) are responsible for generating these unique nanostructures.

Femtosecond lasers can further be used to perform internal modification of glass, as described in Sect. 3.2.4. Internal modification for fabricating functional devices can be realized only by using ultrashort pulse lasers. In 1996, Davis et al.

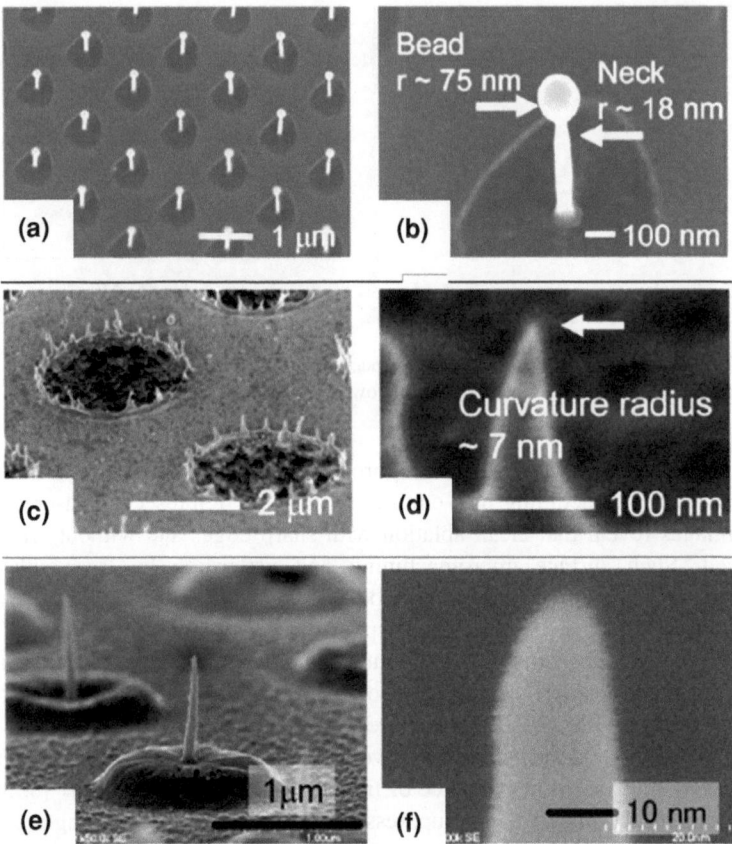

Fig. 3.8 Nanostructures generated on gold thin films. **a** Array of water-drop like nanostrcutures and **b** magnified image of a single water-drop like nanostrcuture on a 10-nm-thick film on a sapphire (0001) substrate. **c** Array of nanocrowns and **d** a spike of a nanocrown on a 50-nm-thick thin film on a silicon (100) substrate. **e** Nanowhiskers and **f** a magnified image of a nanowhisker on a 50-nm-thick film on a sapphire (0001) substrate [34, 35] (Reproduced with permission from SPIE. ©2009, 2011 by SPIE-The International Society for Optics and Photonics)

reported a permanent refractive index change and optical waveguide writing inside glass using a femtosecond laser [13] (see Chap. 6). Many researchers are currently investigating writing of optical waveguides embedded in various glasses such as fused silica, borosilicate glass, and chalcogenide glass. Refractive index modification has been applied to fabricate 3D optical microdevices such as optical couplers and splitters [36], volume Bragg gratings [37], diffractive lenses [38], and distributed feedback (DFB) lasers [39].

Internal modification of glasses by femtosecond laser irradiation modifies their chemical properties as well as their refractive indices. Femtosecond laser direct writing followed by wet chemical etching in dilute HF acid has been used to form

3D hollow microstructures in fused silica. Such microstructures include 3D microfluidic channels with diameters as narrow as 10 μm that can have any interconnection angle between channels and high aspect ratios [40]. Photosensitive glass produced by doping lithium aluminosilicate glass with trace amounts of silver and cerium is an alternative material for this application. It provides a high efficiency and a high processing throughput as well as much smoother etched surfaces, although it requires thermal treatment prior to wet etching [41–43]. Femtosecond laser modification followed by chemical wet etching of fused silica and photosensitive glass is now widely used for fabricating microfluidic and optofluidic devices [44, 45] (see Chaps. 4–6, 8 and 9).

Another important application of femtosecond laser processing is two-photon polymerization (2PP), which is stereolithography (3D lithography) by a near-IR femtosecond laser using epoxy resin. 2PP has a sub-wavelength spatial resolution due to two-photon absorption (see Sect. 3.2.5 and Fig. 3.4). Kawata et al. fabricated a sculpture of the smallest bull in the world by 2PP, as acknowledged by the Guinness Book of Records [17]. It had a spatial resolution in the width direction of 120 nm. Epoxy resin can be replaced with a solid resist to directly fabricate 3D microstructures. 2PP is currently being extensively investigated for fabricating photonic crystals [46–48], micro- and nanosystems [49–51], and lab-on-a-chip (LOC) devices [52, 53] and for medical and tissue engineering [54, 55]. This technique typically has a resolution of about 100 nm; its resolution can be improved to 25 nm by very carefully adjusting the laser power and the scanning speed [56].

Surface texturing of various materials is a crucial area of research in femtosecond laser processing. Various micro and nanoscale textures can be formed by controlling processing parameters such as the beam intensity, the spatial and temporal beam profiles, the wavelength, the polarization, and the processing environment (ambient gas or liquid). The most well-known textured structures are nanoripples, which are formed by femtosecond laser irradiation with a fluence near the ablation threshold [9–12, 57–60]. Surface nanotexturing produced by femtosecond laser irradiation can be applied to reduce the friction between moving components, reduce the adhesive forces of micro and nanocomponents, and increase the adhesion of thin films and medical implants. Other interesting and useful textures that can be formed by femtosecond laser irradiation are regular arrays of conical microstructures; they can be produced on a material surface by irradiating it in a halogen atmosphere (e.g., SF_6 or Cl_2) with hundreds of femtosecond laser pulses [61, 62]. The resulting structures strongly reduce incident light reflection, giving rise to so-called black silicon. They also greatly increase absorption even in the infrared region [63, 64]. This technique is effective for enhancing the efficiency of photovoltaic solar cells [65]. Coating structured surfaces with a layer of silane molecules produces superhydrophobic surfaces due to the lotus effect, which can be used to produce self-cleaning products [62, 66].

3.5 Summary

Femtosecond lasers are promising tools for both surface and volume processing of materials. Reduction of HAZ formation by their ultrashort pulse widths allows high-quality surface micromachining and micropatterning of various kinds of materials to be realized, which have a wide range of practical applications.

Additionally, femtosecond laser irradiation can modify the interior of glass in a spatially selective manner through multiphoton absorption or tunneling ionization. This enables the optical and chemical properties of glass to be simultaneously altered. Optical property modification includes increasing the refractive index in laser-irradiated regions. This can be applied to fabricate 3D photonic microdevices, such as optical waveguides, optical couplers and splitters, volume Bragg gratings, diffractive lenses, and DFB lasers.

The modified chemical properties of laser-irradiated regions in glass allow these regions to be selectively etched by subsequent wet etching using aqueous solutions of etchants such as HF acid. This technique can be used to directly form complex true 3D microfluidic structures. This two-step process can also be used to fabricate free-space optical components such as micromirrors and microlenses inside glass.

The unique ability of femtosecond laser processing to fabricate both microfluidic and optical components in glass opens up new avenues for fabricating a variety of biomicrochips for biological analysis.

References

1. Srinivasan R, Sutcliffe E, Braren B (1987) Ablation and etching of polymethylmethacrylate by very short (160 fs) ultraviolet (308 nm) laser pulses. Appl Phys Lett 51:1285–1287
2. Küper S, Stuke M (1987) Femtosecond uv excimer laser ablation. Appl Phys B 44:99–204
3. Küper S, Stuke M (1989) Ablation of polytetrafluoroethylene (Teflon) with femtosecond UV exicimer laser pulses. Appl Phys Lett 54:4–6
4. Küper S, Stuke M (1989) Ablation of uv-transparent materials with femtosecond UV excimer laser pulses. Microelectron Eng 9:475–480
5. Momma C, Chichkov BN, Nolte S et al (1996) Short-pulse laser ablation of solid targets. Opt Commun 129:134–142
6. Yanik MF, Cinar H, Cinar HN et al (2004) Neurosurgery: functional regeneration after laser axotomy. Nature 432:822–822
7. Barsch N, Korber K, Ostendorf A et al (2003) Ablation and cutting of planar silicon devices using femtosecond laser pulses. Appl Phys A 77:237–242
8. Nakata Y, Okada T, Maeda M (2002) Fabrication of dot matrix, comb, and nanowire structures using laserablation by interfered femtosecond laser beams. Appl Phys Lett 81:4239–4241
9. Reif J, Costache F, Henyk M et al (2002) Ripples revisited: non-classical morphology at the bottom of femtosecond laser ablation craters in transparent dielectrics. Appl Surf Sci 197–198:891–895
10. Wu Q, Ma Y, Fang R et al (2003) Femtosecond laser-induced periodic surface structure on diamond film. Appl Phys Lett 82:1703–1705

11. Rudolph P, Kautek W (2004) Composition influence of non-oxidic ceramics on self-assembled nanostructures due to fs-laser irradiation. Thin Solid Films 453–454:537–541
12. Miyaji G, Miyazaki K (2006) Ultrafast dynamics of periodic nanostructure forma-tion on diamondlikecarbon films irradiated with femtosecond laser pulses. Appl Phys Lett 89:191902
13. Davis KM, Miura K, Sugimoto N et al (1996) Writing waveguides in glass with a femtosecond laser. Opt Lett 21:1729–1731
14. Glezer EN, Milosavljevic M, Huang L et al (1996) Three-dimensional optical storage inside transparent materials. Opt Lett 21:2023–2025
15. Watanabe W, Sowa S, Tamaki T et al (2006) Three-dimensional waveguides fabricated in poly(methyl methacrylate) by a femtosecond laser. Jpn J Appl Phys 45:L765–L767
16. Hanada Y, Sugioka K, Midorikawa K (2010) UV waveguides light fabricated in fluoropolymer CYTOP by femtosecond laser direct writing. Opt Express 18:446–450
17. Kawata S, Sun HB, Tanaka T et al (2001) Finer features for functional microdevices. Nature 412:697–698
18. Fan WS, Storz R, Tom HWK et al (1992) Electron thermalization in gold. Phys Rev B 46:13592–13595
19. Sun CK, Vallée F, Acioli LH et al (1994) Femtosecond-tunable measurement of electron thermalization in gold. Phys Rev B 50:15337–15348
20. Wellershoff SS, Hohlfeld J, Güdde J et al (1999) The role of electron-phonon coupling in femtosecond laser damage of metals. Appl Phys A 69:S99–S107
21. Hohlfeld J, Wellershoff SS, Güdde J et al (2000) Electron and lattice dynamics following optical excitation of metals. Chem Phys 251:237–258
22. Anisimov SI, Rethfeld B (1997) Theory of ultrashort laser pulse interaction with a metal. Proc SPIE 3093:192–203
23. Corkum PB, Brunel F, Sherman NK et al (1988) Thermal response of metals to ultrashort pulse laser excitation. Phys Rev Lett 61:2886–2889
24. Fujita M, Hashida M (2004) Applications of femtosecond lasers. Oyo Buturi 73:178–185 (in Japanese)
25. Keldysh LV (1965) Ionization in field of a strong electromagnetic wave. Sov Phys JETP 20:1307–1314
26. Mao SS, Quere F, Guizard S et al (2004) Dynamics of femtosecond laser interactions with dielectrics. Appl Phys A 79:1695–1709
27. Chichkov BN, Momma C, Nolte S et al (1996) Femtosecond, picosecond and nanosecond laser ablation of solids. Appl Phys A 63:109–115
28. Nakashima S, Sugioka K, Midorikawa K (2010) Enhancement of resolution and quality of nano-hole structure on GaN substrates using the second-harmonic beam of near-infrared femtosecond laser. Appl Phys A 101:475–481
29. Nakashima S, Sugioka K, Ito T et al (2011) Fabrication of high-aspect-ratio nanohole arrays on GaN surface by using wet-chemical-assisted femtosecond laser ablation. J Laser Micro/ Nanoeng 6:15–19
30. Nakata Y, Okada T, Maeda M (2003) Nano-sized hollow bump array generated by single femtosecond laser pulse. Jpn J Appl Phys 42:L1452–L1454
31. Nakata Y, Okada T, Maeda M (2004) Lithographical laser ablation using femto-second laser. Appl Phys A 79:1481–1483
32. Nakata Y, Miyanaga N, Okada T (2007) Effect of pulse width and fluence of femtosecond laser on the size of nanobump array. Appl Surf Sci 253:6555–6557
33. Nakata Y, Tsuchida K, Miyanaga N et al (2009) Liquidly process in femtosecond laser processing. Appl Surf Sci 255:9761–9763
34. Nakata Y, Hiromoto T, Miyanaga N (2009) Frozen water drops in the nanoworld. SPIE Newsroom. doi:10.1117/2.1200906.1708
35. Nakata Y, Momoo K, Hiromoto T et al (2011) Generation of superfine structure smaller than 10 nm by interfering femtosecond laser processing. Proc SPIE 7920:79200B
36. Watanabe W, Asano T, Yamada K et al (2003) Wavelength division with three-dimensional couplers fabricated by filamentation of femtosecond laser pulses. Opt Lett 28:2491–3493

37. Sudrie L, Winick KA (2003) Fabrication and characterization of photonic devices directly written in glass using femtosecond laser pulses. J Lightwave Technol 21:246–253
38. Bricchi E, Mills JD, Kazamsky PG et al (2002) Birefringent Fresnel zone plates in silica fabricated by femtosecond laser machining. Opt Lett 27:2200–2202
39. Valle GD, Taccheo S, Osellame R et al (2007) 1.5μm single longitudinal mode waveguide laser fabricated by femtosecond laser writing. Opt Express 84:3190–3194
40. Marcinkevicius A, Juodkazis S, Watanabe M et al (2001) Femtosecond laser-assisted three-dimensional microfabrication in silica. Opt Lett 26:277–279
41. Masuda M, Sugioka K, Cheng Y et al (2003) 3-D microstructuring inside photosensitive glass by femtosecond laser excitation. Appl Phys A 76:857–860
42. Sugioka K, Cheng Y, Midorikawa K (2005) Three-dimensional micromachining of glass using femtosecond laser for lab-on-a-chip device manufacture. Appl Phys A 81:1–10
43. Sugioka K, Hanada Y, Midorikawa K (2010) Three-dimensional femtosecond laser micromachining of photosensitive glass for biomicrochips. Laser Photon Rev 4:386–400
44. Sugioka K, Cheng Y (2011) Integrated microchips for biological analysis fabricated by femtosecond laser direct writing. MRS Bull 36:1020–1027
45. Sugioka K, Cheng Y (2011) Femtosecond laser processing for optofluidic fabrication. Lab Chip 12:3576–3589
46. Cumpston B, Ananthavel S, Barlow S et al (1999) Two-photon polymerization initiators for three-dimensional optical data storage and microfabrication. Nature 398:51–54
47. Sun HB, Matsuo S, Misawa H (1999) Three-dimensional photonic crystal structures achieved with two-photon-absorption photopolymerization of resin. Appl Phys Lett 74:786–788
48. Serbin J, Ovsianikov A, Chichkov B (2004) Fabrication of woodpile structures by two-photon polymerization and investigation of their optical properties. Opt Express 12:5221–5228
49. Maruo S, Inoue H (2006) Optically driven micropump produced by three-dimensional two-photon microfabrication. Appl Phys Lett 89:144101
50. Maruo S, Inoue H (2007) Optically driven viscous micropump using a rotating microdisk. Appl Phys Lett 91:084101
51. Tian Y, Zhang YL, Ku JF et al (2010) High performance magnetically controllable microturbines. Lab Chip 10:2902–2905
52. Wang J, He Y, Xia H et al (2010) Embellishment of microfluidic devices via femtosecond laser micronanofabrication for chip functionalization. Lab Chip 10:1993–1996
53. Wu D, Chen QD, Niu LG et al (2009) Femtosecond laser rapid prototyping of nanoshells and suspending components towards microfluidic devices. Lab Chip 9:2391–2394
54. Ovsianikov A, Malinauskas M, Schlie S et al (2011) Three-dimensional laser micro- and nano-structuring of acrylated poly(ethylene glycol) materials and evaluation of their cytoxicity for tissue engineering applications. Acta Biomater 7:967–974
55. Farsari M, Chichkov B (2009) Two-photon fabrication. Nature Photon 3:450–452
56. Tan DF, Li Y, Qi FG et al (2007) Reduction in feature size of two-photon polymerization using SCR500. Appl Phys Lett 90:071106
57. Sakabe S, Hashida M, Tokita S et al (2009) Mechanism for self-formation of periodic grating structures on a metal surface by a femtosecond laser pulse. Phys Rev B 79:033409
58. Yasumaru N, Miyazaki K, Kiuchi J (2003) Femtosecond-laser-induced nanostructure formed on hard thin films of TiN and DLC. Appl Phys A 76:983–985
59. Borowiec A, Hauge HK (2003) Subwavelength ripple formation on the surfaces of compound semiconductors irradiated with femtosecond laser pulses. Appl Phys Lett 82:4462–4464
60. Costache F, Henyk M, Reif J (2003) Surface patterning on insulators upon femtosecond laser ablation. Appl Surf Sci 208:486–491
61. Her TH, Finlay RJ, Wu C et al (1998) Microstructuring of silicon with femtosecond laser pulses. Appl Phys Lett 73:1673–1675
62. Baldacchini T, Carey JE, Zhou M et al (2006) Superhydrophobic surfaces prepared by microstructuring of silicon using a femtosecond laser. Langmuir 22:4917–4919
63. Carey JE, Crouch CH, Shen M et al (2005) Visible and near-infrared responsivity of femtosecond-laser microstructured silicon photodiodes. Opt Lett 30:1773–1775

64. Younkin R, Carey JE, Mazur E et al (2003) Infrared absorption by conical silicon microstructures made in a variety of background gases using femtosecond-laser pulses. J Appl Phys 93:2626–2629
65. Wang F, Chen C, He H et al (2011) Analysis of sunlight loss for femtosecond laser microstructed silicon and its solar cell efficiency. Appl Phys A 103:977–982
66. Zorba V, Stratakis E, Barberoglou M et al (2008) Biomimetic artificial surfaces quantitatively reproduce the water repellency of a lotus leaf. Adv Mater 20:4049–4054

Chapter 4
Fabrication of Microfluidic Structures in Glass

Abstract Although laser drilling has long been used for producing straight one-dimensional (1D) holes in glass, it generally cannot be used to form 3D microchannels since thin channels become clogged with the debris produced during laser ablation. This chapter describes two approaches that have been developed to overcome this problem. The first is femtosecond-laser-assisted wet chemical etching, in which femtosecond laser irradiation is used to modify the chemical properties of glass and subsequent chemical etching is used to selectively remove the modified regions. The second approach is liquid-assisted femtosecond laser 3D drilling in which liquid is flowed through the channels to greatly enhance the removal rate of debris produced by laser ablation. This chapter also discusses several beam-shaping techniques for controlling the cross section of the microchannels. The cross-sectional shape of microchannels is significant in many microfluidic applications because it determines the fluid dynamics and biological functions of microchannels.

4.1 Introduction

Microfluidic devices have undergone rapid development in the past two decades, enabling systems for chemical and biological analysis to be miniaturized [1, 2]. A wide variety of microfluidic devices have been fabricated for controlling and manipulating tiny volumes of liquids with high precision and ease of operation. The most popular microfluidic fabrication technology is soft lithography using poly (dimethylsiloxane) (PDMS) substrates [3] (see Chap. 2). Although soft lithography is rapid and cost effective, it cannot be used to directly form three-dimensional (3D) microfluidic structures such as buried microchannels and microchambers without stacking and bonding. This difficulty can be overcome using the advanced femtosecond laser 3D microprocessing techniques introduced in Chap. 1. This chapter provides an overview of state-of-the-art fabrication techniques for microfluidic applications.

K. Sugioka and Y. Cheng, *Femtosecond Laser 3D Micromachining for Microfluidic and Optofluidic Applications*, SpringerBriefs in Applied Sciences and Technology, DOI: 10.1007/978-1-4471-5541-6_4, © The Author(s) 2014

The remainder of the chapter is organized as follows. Section 4.2 describes the fabrication of microfluidic structures in substrates consisting of photosensitive glass Foturan and fused silica using femtosecond-laser-assisted wet chemical etching. Section 4.3 describes fabrication of microfluidic structures in fused silica using liquid-assisted femtosecond laser 3D drilling. Finally, Sect. 4.4 describes several beam shaping techniques for controlling the cross-sectional shapes of microchannels.

4.2 Femtosecond-Laser-Assisted Wet Chemical Etching

4.2.1 Fabrication in Photosensitive Glass

Photosensitive glass that can react with UV light was first discovered by S. Donald Stookey over half a century ago [4]. Currently, the most widely used photosensitive glass is Foturan, which is manufactured by Schott Glass Corp. Foturan glass is conventionally structured by single-photon absorption of UV light with a wavelength shorter than 320 nm. The mechanism of single-photon induced photostructuring of Foturan glass is well understood. Foturan glass is lithium aluminosilicate glass doped with trace amounts of silver and cerium. The cerium (Ce^{3+}) ions act as a photosensitizer: when a Ce^{3+} ion absorbs a UV photon, it releases an electron and becomes Ce^{4+}. The silver ions in Foturan glass then capture some of the free electrons and become silver atoms. In the subsequent heat treatment, these silver atoms diffuse and agglomerate to form nanoclusters at ~ 500 °C. The annealing temperature is raised to ~ 600 °C to promote growth of the crystalline phase of lithium metasilicate in the amorphous glass matrix using the silver clusters as nuclei [5]. Since the crystalline phase of lithium metasilicate has a ~ 50 times higher etch rate than the unmodified glass matrix, it can be preferentially etched in dilute HF acid, leaving 2D structures on the surfaces of Foturan glass. UV lamps are conventionally used as the illumination source for 2D microfabrication on Foturan glass surfaces.

3D fabrication can be realized using nonlinear multiphoton absorption in Foturan instead of linear single-photon absorption since this enables internal modification if it is combined with laser direct writing. Fabrication of microfluidic structures directly embedded in Foturan glass was first demonstrated using ultraviolet (UV) nanosecond laser direct writing and subsequent wet chemical etching in hydrofluoric (HF) etchant in 1997 by Helvajian's group [6]. Since the UV laser they used had a wavelength of 355 nm, which is too long to realize single-photon absorption in Foturan, cerium ions were photoionized by spatially localized nonlinear two-photon absorption that occurred only in the vicinity of the focal volume (this is similar to two-photon polymerization described in Chap. 2). Later, in 1999, Hirao's group fabricated a Y-shaped microfluidic channel structure in a home-produced photosensitive glass that had a similar composition to Foturan using a frequency-doubled femtosecond laser operated at a wavelength of ~400 nm [7].

Near-IR (i.e., wavelengths longer than ~800 nm) femtosecond lasers with various pulse durations and repetition rates are currently the dominant light source for fabricating 3D microstructures embedded in photosensitive glass [8–10]. In multiphoton absorption, this relatively long wavelength gives a high nonlinearity in the photoreaction, enhancing the axial resolution. This is vital for achieving true 3D microprocessing. The photoreaction initiated by near-IR femtosecond laser irradiation has a different mechanism from that of single- or two-photon excitation by UV irradiation because the interaction of high-intensity femtosecond laser pulses with Foturan glass can directly produce a large amount of free electrons even without a photosensitizer. Consequently, it is not necessary to dope Foturan glass with cerium provided femtosecond lasers are used [11].

Figure 4.1 schematically illustrates the process used to fabricate 3D microfluidic structures in Foturan glass: formation of a latent image by femtosecond laser direct writing (Fig. 4.1a), transformation of the latent image into an etchable phase by thermal treatment (Fig. 4.1b), and removal of the modified material by wet chemical etching in a 5–10 % aqueous solution of HF acid in an ultrasonic bath (Fig. 4.1c). The ultrasonic bath is critical because it can significantly enhance the etch rate by simply increasing the mass transfer of the chemical etchant in the thin channel.

Figure 4.2a and b respectively show micrographs of the glass substrate surface and microchannels embedded in glass of a typical microfluidic mixer fabricated in Foturan glass [12]. The two microreservoirs on the right-hand side serve as solution inlets while the horizontal hole on the left-hand side serves as the solution outlet. To perform mixing, two different liquids are introduced into the micromixer

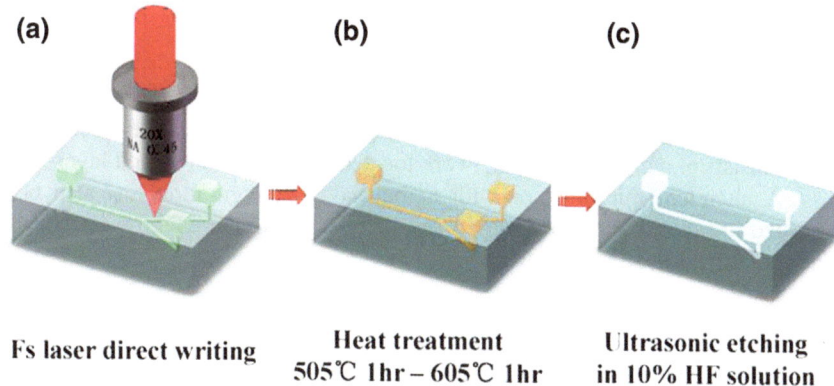

(a) Fs laser direct writing

(b) Heat treatment
505 ℃ 1hr – 605 ℃ 1hr

(c) Ultrasonic etching
in 10% HF solution

Fig. 4.1 Schematic depiction of the processes for fabricating 3D microfluidic structures in Foturan glass: **a** formation of a latent image by femtosecond laser direct writing; **b** transformation of the latent image into an etchable phase by thermal treatment; and **c** removal of the modified material by wet chemical etching in a 5–10 % aqueous solution of HF acid in an ultrasonic bath. The ultrasonic bath is critical because it can significantly enhance the etch rate by simply promoting mass transfer of the chemical etchant in the thin channel

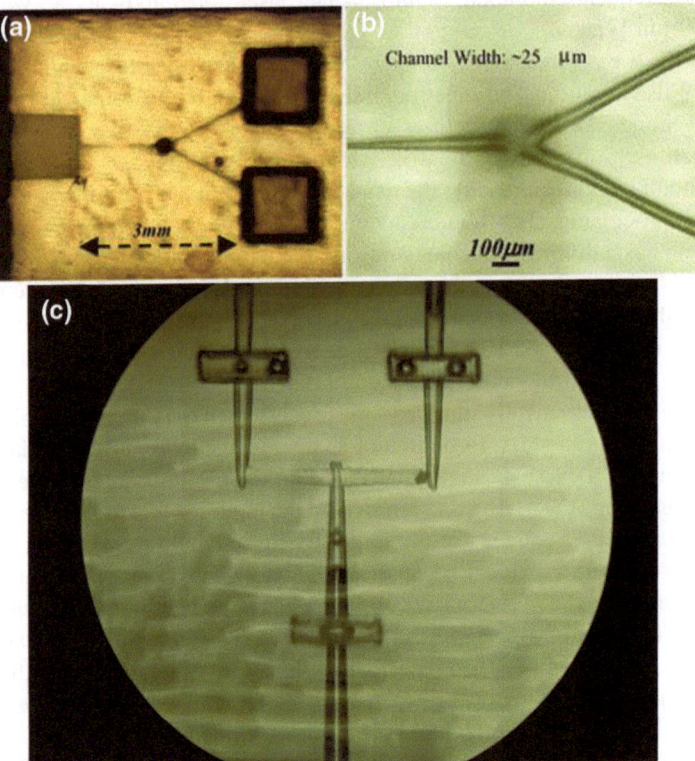

Fig. 4.2 Optical micrographs showing (**a**) *top* view of a micromixer, and (**b**) *close-up* view of microchannels embedded in glass [12]. **c** Optical micrograph of a complex 3D microfluidic structure with microchannels and chambers arranged in a multilayer configuration [13]

from the two inlets. Both solutions are transferred to the mixing channel by reducing the pressure in a silicon tube connected to the outlet. The microfluidic mixer consists of a microchannel network located in a single horizontal layer in the glass. A chemical microreactor with a multilayer configuration has also been fabricated in Foturan glass (see Fig. 4.2c), demonstrating the effectiveness of this technique for fabricating 3D structures [13].

4.2.2 Fabrication in Fused Silica

It is reasonable to expect Foturan glass to be photoetchable by femtosecond laser pulses as a multiphoton counterpart of single-photon UV lithography. Interestingly, photoetching can also be induced in fused silica by femtosecond laser irradiation, despite fused silica being conventionally regarded as a non-

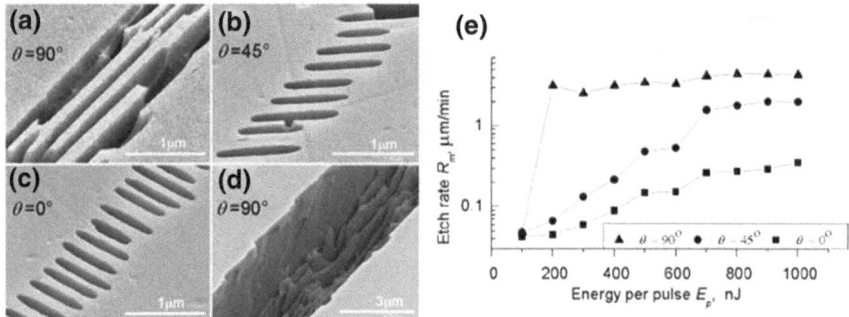

Fig. 4.3 **a–c** *Top*-view (*xy*-plane) SEM images of long-range periodic nanostructures formed along the writing direction for different polarizations for $E_p = 300$ nJ (the polarization is parallel to the writing direction when $\theta = 0°$), and (**d**) $E_p = 900$ nJ. The structures were revealed after etching for 20 min in a 0.5 % aqueous solution of HF. **e** Etch rate of femtosecond-laser-modified zones as a function of pulse energy for three different polarizations of the writing beam [16] (Reproduced with permission from OSA. ©2005 by the Optical Society of America)

photosensitive material in the visible region. Misawa's group pioneered fabrication of 3D microfluidics in fused silica using a femtosecond laser in 2001, although they obtained a low etch selectivity [14]. Later in 2004, Bellouard et al. formed high-aspect ratio microchannels with arbitrary lengths that are open on the top surface of fused silica by femtosecond-laser-assisted wet chemical etching [15].

A substantial breakthrough in fabrication in fused silica was achieved in 2005 when Hnatovsky et al. discovered that the etch rate and selectivity strongly depend on the polarization of the writing pulses [16]. This is due to the formation of periodic nanograting-like structures that always oriented perpendicular to the polarization of the laser beam [17, 18]. As shown in Fig. 4.3a–d, these nanogratings consist of alternating regions with high and low etch rates. If the nanogratings are oriented perpendicular to the microchannel (Fig. 4.3c), the alternating regions will prevent the HF solution from entering the microchannel, resulting in the lowest etch rate. Therefore, to achieve the highest etch rate, the nanogratings must be oriented parallel to the microchannel; this has the potential to enhance the etch rate by nearly two orders of magnitude relative to that obtained when the nanogratings are oriented perpendicular to the microchannel, as shown in Fig. 4.3e.

Further enhancement of the etch selectivity (but not the etch rate) has been achieved using KOH solution instead of dilute HF acid solution as the etchant, as shown in Fig. 4.4 [19]. It clearly shows that the channels formed using KOH have much more uniform diameters. Kiyama et al. recently used a KOH etchant to fabricate a 1-cm-long through channel that had a diameter of less than 60 μm near its ends and a much smaller diameter at its middle. Since it was produced without using an ultrasonic bath, it may be possible to further improve the microchannel homogeneity.

Since the etch selectivity cannot be increased indefinitely, microchannels fabricated by femtosecond-laser-assisted chemical etching will inevitably exhibit

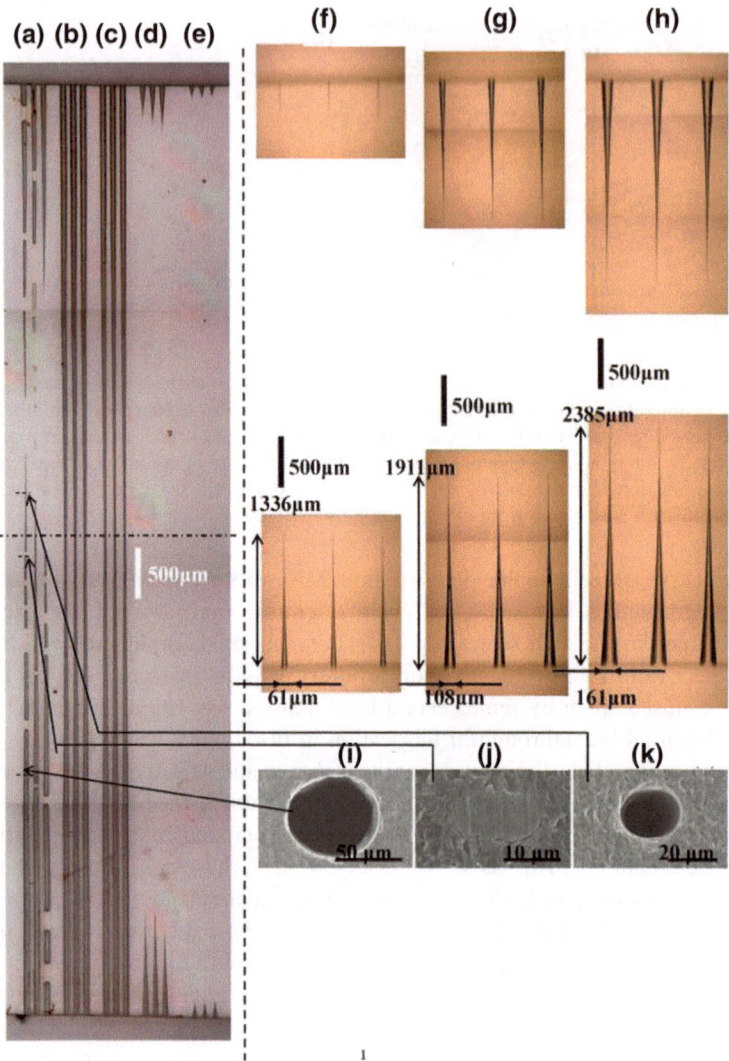

Fig. 4.4 Comparison of etching profiles in a 9.2-mm-long silica substrate (9.2 × 10 × 0.5 mm) produced by etching by 10 M (35.8 %) aqueous KOH (20 mL) and 2.0 % aqueous HF (20 mL) observed by reflectance optical microscopy. **a–e** Channels fabricated after soaking for 60 h in KOH at 80 °C after irradiation by femtosecond pulse trains (interval: 0.1 μm) focused by a 40 × objective lens (NA: 0.65) at a depth of 10 μm below the surface for pulse energies of (**a**) 500, (**b**) 400, (**c**) 300, (**d**) 200, and (**e**) 100 nJ. **f–h** Two images of each set of etching fronts, which represent the time evolution of channel formation in aqueous HF solution at ambient temperature. A sample that had been irradiated by pulses with energies of 360 nJ through a 40 × objective lens (NA: 0.65) was submerged in the solution for (**f**) 24, (**g**) 48, and (**h**) 72 h. Pulse trains were irradiated from top to bottom with the electric vector of the laser beam oriented parallel to the scanning direction. Laser writing was performed three times at each laser pulse energy to confirm reproducibility. **i–k** Cross-sectional field emission SEM images of the channel on the *far left* in (**a**), showing open holes (**i**) and (**k**) and a hole filled with precipitates (**j**) [19] (Reprinted with permission from ACS. Copyright 2009 American Chemical Society)

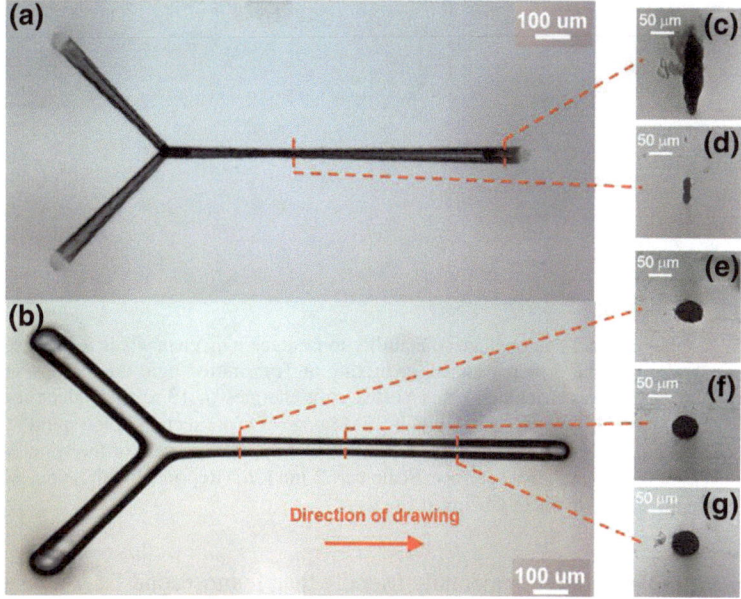

Fig. 4.5 Optical micrographs of Y-branched channel. *Top* view of channel (**a**) before, (**b**) after drawing, and (**c–g**) cross-sectional views of channel at locations indicated by *dashed lines* [21]

tapering, which will induce variations in the fluid flow velocity. One approach to produce homogeneous microchannels is to pre-compensate for the tapering by suitably wobbling the glass sample during laser irradiation [20]; however, this strategy can only be used when fabricating wide channels and the channel length is limited to a few millimeters. To fabricate narrow channels with a high uniformity and a long length, glass drawing can be applied to a glass sample containing a tapered microfluidic channel [21]. As shown in Fig. 4.5, after drawing, the microchannel cross section becomes perfectly symmetrical and the channel diameter becomes uniform along the entire channel length. This technique allows narrow microfluidic channels to be produced with long lengths (e.g., an aspect ratio higher than ~1000). Glass drawing also produces an extremely smooth inner surface with an RMS roughness of ~0.3 nm, which is beneficial for optical and optofluidic applications.

4.3 Liquid-Assisted Femtosecond Laser Three-Dimensional Drilling

Material removal from laser-treated areas can be significantly enhanced by performing femtosecond laser 3D drilling in distilled water from the rear surface of the glass, as first demonstrated by Li et al. [22]. This technique is easier to

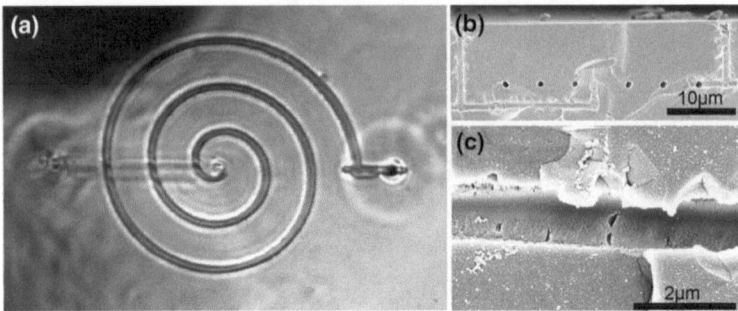

Fig. 4.6 A spiral channel demonstrates the ability to produce long channels in a small space for separation, hydrodynamic flow resistance, or mixing. **a** Transmitted light microscopy image of the spiral. The spiral was machined using pulses with energies of 18 nJ to produce a channel diameter of 900 nm and a length of 143 μm. **b** SEM image of *cross section* of the spiral revealing that it does not contain debris. Scale bar: 10 μm. **c** *Close up* of a segment of the spiral showing the roughness of the inner channel surface. Scale bar: 2 μm [26] (Reprinted with permission from ACS. Copyright 2005 American Chemical Society)

implement and more environmentally friendly than femtosecond-laser-assisted wet chemical etching. More importantly, since it does not rely on generating etch selectivity in materials by femtosecond laser irradiation, it can be applied to any material that is transparent to the writing pulses [23]. It can also provide a more uniform channel diameter for the same reason. Its main problem is that when the drilling length reaches several hundreds of micrometers, debris generated by femtosecond laser ablation can still clog the microchannel, restricting the size of fabricated microstructures to ∼1 mm [24, 25].

An important feature of liquid-assisted femtosecond laser drilling is that it can be used to produce very narrow channels inside glass, as shown in Fig. 4.6 [26]. These subsurface nanochannels, which have diameters of only ∼700 nm and can have arbitrary geometries, were fabricated using low-energy (i.e., near ablation threshold) femtosecond laser pulses tightly focused by a high-numerical-aperture (NA) objective lens (see Fig. 3.4). The channel lengths are again limited to several hundred micrometers because debris cannot be expelled from longer channels by bubbles created in water by laser irradiation.

A new strategy has recently been developed for fabricating microchannels with nearly unlimited lengths and arbitrary geometries by femtosecond laser direct writing in mesoporous glass immersed in water followed by post-annealing, realizing long square-wave-shaped microchannels [27] and large-volume micro-fluidic chambers [28]. Figure 4.7a and b respectively show a schematic view of the experimental setup and a flow diagram of the fabrication process. The main fabrication process involves two steps: (1) direct formation of hollow microchannels in the porous glass substrate immersed in water by femtosecond laser ablation and (2) post-annealing at ∼1150 °C to consolidate the porous glass substrate. As a demonstration, a square-wave-like microchannel with a total length of ∼14 mm

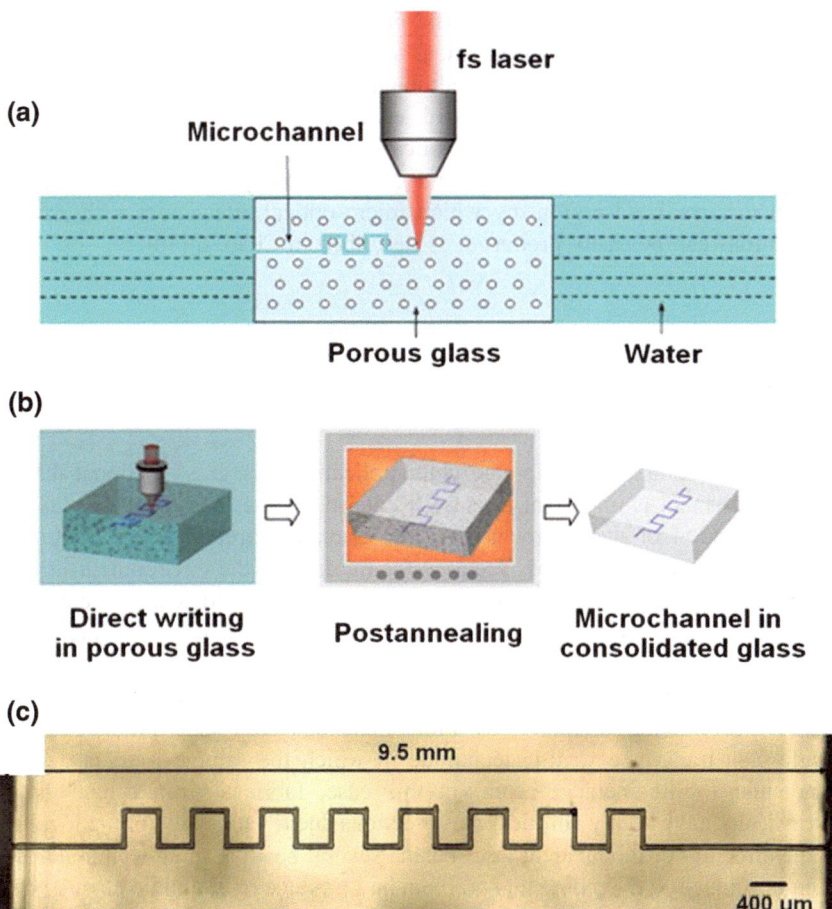

Fig. 4.7 a Schematic diagram of experimental setup, and (**b**) flow diagram of fabrication process. **c** Optical micrograph of a square-wave-like microchannel fabricated in glass [27]

and a diameter of ∼64 μm has been fabricated at a depth of ∼250 μm below the glass surface (see Fig. 4.7c). The channel diameter has an excellent homogeneity. Using this technique, a passive microfluidic mixer consisting of 3D mixing units has been constructed. The 3D micromixer has a higher mixing efficiency than a 1D micromixer [29] (see also Chap. 9).

Currently, the most common glass materials used for fabricating 3D micro-fluidic structures are Foturan glass and fused silica. They both have their own niche applications. Fused silica exhibits better optical properties including a broader transmission range and lower autofluorescence, which are desirable for optofluidic applications for which a superior optical performance is of first importance. On the other hand, it is more advantageous to use Foturan glass to fabricate large-scale or complex microfluidic structures due to two factors.

First, since it is a photosensitive glass, Foturan glass can be modified at much lower peak intensities than fused silica, resulting in higher scanning speeds and hence lower writing times. Second, although Foturan glass and fused silica have comparable femtosecond-laser-induced etch selectivities, Foturan provides a higher overall etch rate than fused silica. Foturan glass also permits selective precipitation of silver nanoparticles and growth of a crystalline lithium metasilicate in the irradiated region, by which micro-optical attenuators with arbitrary attenuations can be easily produced, as described in Chap. 6 [30].

4.4 Shaping Techniques for Fabricating Microfluidic Channels

For many microfluidic and lab-on-a-chip applications, the cross-sectional shape of the microchannel is important because it determines the fluid dynamics and biological function of the microchannel. For example, circular cross sections are preferable to rectangular cross sections for mimicking the environment in blood vessels since rectangular cross sections induce non-physiological gradients in the fluid shear rate, velocity, and pressure. The cross-sectional shapes of microfluidic channels with large cross sections (i.e., significantly larger than the femtosecond laser wavelength) can be tailored by overlapping multiple laser-affected zones [15]. However, small-diameter microfluidic channels are usually produced using a single-scan transverse writing technique in which the sample is translated perpendicular to the incident beam. In this case, fabricated microchannels will intrinsically have highly elliptical cross sections due to elongation of the focal spot in the direction of the incident laser beam. Several beam shaping techniques have been developed to overcome this problem and produce microfluidic channels with circular cross sections. Typical shaping techniques include astigmatic beam shaping [31], slit beam shaping [32], crossed-beam shaping [33], and spatiotemporal beam shaping [34, 35].

Figure 4.8a schematically illustrates a focusing system that incorporates a slit beam shaping technique [32]. It employs a narrow slit that can be placed directly in front of the objective lens without using an extra mount. The diffraction induced by the slit enlarges the focal spot in the transverse direction, producing a microchannel with a more symmetrical cross section (Fig. 4.8b) than a microchannel fabricated without slit beam shaping (Fig. 4.8c). It is important that the slit be oriented parallel to the direction of sample motion in the laser writing process, but this makes it difficult to fabricate curved microchannels with complex 3D geometries. This problem can be solved by simultaneous shaping the femtosecond laser pulses both spatially and temporally [34].

A 3D isotropic resolution can be obtained using a crossed-beam shaping technique, as schematically shown in Fig. 4.9a [33]. The focusing system consists of two orthogonal objective lenses that are positioned so that they have a common

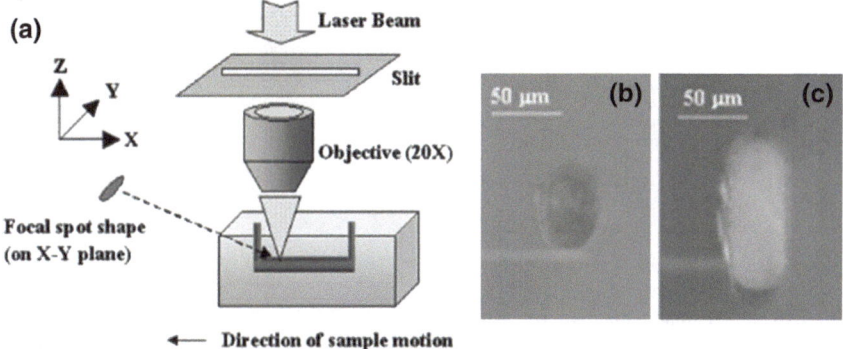

Fig. 4.8 **a** Schematic diagram of focusing system incorporating slit beam shaping technique, and *cross-sectional* views of microfluidic channels fabricated, **b** with, and **c** without the use of slit beam shaping technique [32]

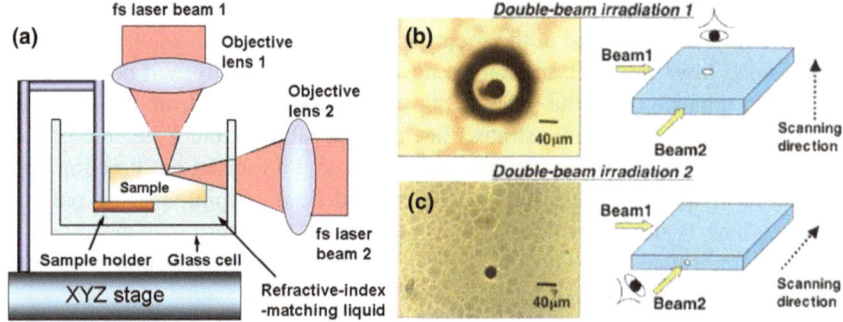

Fig. 4.9 **a** Schematic diagram of focusing system incorporating crossed-beam shaping technique and cross-sectional views of microfluidic channels fabricated by translating the sample, **b** perpendicular to the plane formed by the crossed beams, and **c** parallel to one of the two beams [33]

focal point. If two femtosecond laser pulses temporally overlap at the common focal point, the synthesized focal spot will produce an isotropic illumination volume near the crossing point. Figure 4.9b shows that circular cross sections can be obtained for samples fabricated with both horizontal and vertical translation directions. The major difficulty with this technique is maintaining the spatial and temporal overlap of the two femtosecond laser beams focused by the orthogonal objective lenses during the fabrication process. This difficulty can be overcome by translating the glass in a refractive-index-matching liquid.

It has recently been demonstrated that a 3D isotropic resolution can also be achieved with a spatiotemporal pulse shaping system in which the incident pulses are first spatially dispersed by a pair of parallel gratings before entering the

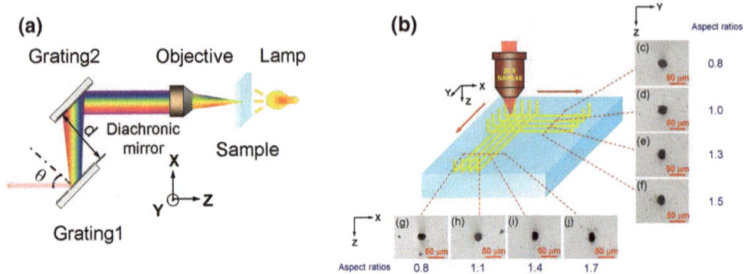

Fig. 4.10 a Schematic diagram of spatiotemporal focusing system, and **b** 3D microfluidic channels fabricated using the temporal focusing method. The *red arrows* indicate the translation directions of the stage. **c–j** Optical micrographs of *cross sections* of microfluidic channels fabricated with different beam sizes and laser powers [34]

focusing lens, as schematically illustrated in Fig. 4.10a [34]. Temporal focusing is achieved by focusing an objective lens because different frequency components spatially overlap only near the focus. Hence, the pulse width is minimized and the peak intensity is maximized at the focus. This will improve the axial resolution of femtosecond laser microfabrication because the peak intensity will decrease rapidly due to broadening of the pulse width when moving away from the geometric focal spot. Figure 10b demonstrates that it is possible to fabricate microfluidic channels with circular cross sections regardless of the translation direction. Spatiotemporal shaping is much simpler than crossed-beam shaping because only one objective lens is used and hence it is not difficult to ensure that the pulses of multiple femtosecond laser beams spatially and temporally overlap at a common focal point.

4.5 Summary

Fabrication of 3D microfluidic structures in glass has been mainly achieved by employing two technical approaches: femtosecond-laser-assisted wet chemical etching and liquid-assisted femtosecond laser 3D drilling. The most common glass materials used in these applications are photosensitive Foturan glass and fused silica. They both have their own niche applications due to their unique optical and chemical properties. Previously, microchannels fabricated by both femtosecond-laser-assisted wet chemical etching and liquid-assisted femtosecond laser 3D drilling were limited to lengths of less than ∼1 cm. Recently, microfluidic channels with nearly unlimited lengths and arbitrary 3D geometries have been created by liquid-assisted femtosecond laser 3D drilling of porous glass, followed by high-temperature post-annealing to collapse all the nanopores. Without this post-annealing, microchannels fabricated in porous glass are leaky due to the presence of interconnected nanopores around the microchannels.

For microfluidic applications, it is important to control the cross section of microchannels. This can be achieved by spatially and/or temporally shaping the writing beams. Of the above-mentioned techniques, the slit beam shaping technique has received most attention for writing straight 1D channels because it is effective and extremely easy to implement. However, the spatiotemporal shaping technique can provide a truly isotropic 3D resolution using only one focusing lens, making it attractive for fabricating complex 3D microfluidic circuits in glass. Shaping techniques can also be used for writing optical waveguides with circular cross sections [31, 36–38], which is critical to ensure single-mode propagation, as is described in detail in Chap. 6.

References

1. Manz A, Graber N, Widmer HM (1990) Miniaturized total chemical analysis systems: a novel concept for chemical sensing. Sens Actuators, B 1:244–248
2. Whitesides GM (2006) The origins and the future of microfluidics. Nature 442:368–373
3. McDonald JC, Whitesides GM (2002) Poly (dimethylsiloxane) as a material for fabricating microfluidic devices. Acc Chem Res 35:491–499
4. Stookey SD (1950) Photosensitive gold glass and method of making it. US Pat No 2515937, 18 Jul 1950
5. Fuqua P, Janson SW, Hansen WW et al (1999) Fabrication of true 3D microstructures in glass/ceramic materials by pulsed UV laser volumetric exposure techniques. Proc SPIE 3618:213–220
6. Hansen WW, Janson SW, Helvajian H (1997) Direct-write UV-laser microfabrication of 3D structures in lithium-aluminosilicate glass. Proc SPIE 2991:104–112
7. Kondo Y, Qiu JR, Mitsuyu T et al (1999) Three-dimensional microdrilling of glass by multiphoton process and chemical etching. J Jpn Appl Phys 38:L1146–L1148
8. Masuda M, Sugioka K, Cheng Y et al (2003) 3-D microstructuring inside photosensitive glass by femtosecond laser excitation. Appl Phys A 76.857–860
9. Cheng Y, Sugioka K, Masuda M et al (2003) 3D microstructuring inside Foturan glass by femtosecond laser. RIKEN Rev 50:101–106
10. Sugioka K, Cheng Y (2011) Integrated microchips for biological analysis fabricated by femtosecond laser direct writing. MRS Bull 36:1020–1027
11. Hongo T, Sugioka K, Niino H et al (2005) Investigation of photoreaction mechanism of photosensitive glass by femtosecond laser. J Appl Phys 97:063517(4)
12. Cheng Y, Sugioka K, Midorikawa K et al (2005) Microfabrication of 3D hollow structures embedded in glass by femtosecond laser for Lab-on-a-chip applications. Appl Surf Sci 248:172–176
13. Sugioka K, Masuda M, Hongo T et al (2004) Three-dimensional microfluidic structure embedded in photostructurable glass by femtosecond laser for lab-on-chip application. Appl Phys A 78:815–817
14. Marcinkevičius A, Juodkazis S, Watanabe M et al (2001) Femtosecond laser-assisted three-dimensional microfabrication in silica. Opt Lett 26:277–279
15. Bellouard Y, Said A, Dugan M et al (2004) Fabrication of high-aspect ratio, micro-fluidic channels and tunnels using femtosecond laser pulses and chemical etching. Opt Express 12:2120–2129
16. Hnatovsky C, Taylor RS, Simova E et al (2005) Polarization-selective etching in femtosecond laser-assisted microfluidic channel fabrication in fused silica. Opt Lett 30:1867–1869

17. Shimotsuma Y, Kazansky PG, Qiu J et al (2003) Self-organized nanogratings in glass irradiated by ultrashort light pulses. Phys Rev Lett 91:247405(4)
18. Bhardwaj VR, Simova E, Rajeev PP et al (2006) Optically produced arrays of nano-planes inside fused silica. Phys Rev Lett 96:057404(4)
19. Kiyama S, Matsuo S, Hashimoto S et al (2009) Examination of etching agent and etching mechanism on femtosecond laser microfabrication of channels inside vitreous silica substrates. J Phys Chem C 113:11560–11566
20. Vishnubhatla KC, Bellini N, Ramponi R et al (2009) Shape control of microchannels fabricated in fused silica by femtosecond laser irradiation and chemical etching. Opt Express 17:8685–8695
21. He F, Cheng Y, Xu Z et al (2010) Direct fabrication of homogeneous microfluidic channels embedded in fused silica using a femtosecond laser. Opt Lett 35:282–284
22. Li Y, Itoh K, Watanabe W et al (2001) Three-dimensional hole drilling of silica glass from the rear surface with femtosecond laser pulses. Opt Lett 26:1912–1914
23. Kim TN, Campbell K, Groisman A et al (2005) Femtosecond laser-drilled capillary integrated into a microfluidic device. Appl Phys Lett 86:201106(3)
24. An R, Li Y, Dou Y et al (2005) Simultaneous multi-microhole drilling of soda-lime glass by water-assisted ablation with femtosecond laser pulses. Opt Express 13:1855–1859
25. Hwang DJ, Choi TY, Grigoropoulos CP et al (2004) Liquid-assisted femtosecond laser drilling of straight and three-dimensional microchannels in glass. Appl Phys A 79:605–612
26. Ke K, Hasselbrink EF Jr, Hunt AJ et al (2005) Rapidly prototyped three-dimensional nanofluidic channel networks in glass substrates. Anal Chem 77:5083–5088
27. Liao Y, Ju Y, Zhang L et al (2010) Three-dimensional microfluidic channel with arbitrary length and configuration fabricated inside glass by femtosecond laser direct writing. Opt Lett 35:3225–3227
28. Ju Y, Liao Y, Zhang L et al (2012) Fabrication of large-volume microfluidic chamber embedded in glass using three-dimensional femtosecond laser micromachining. Microfluid Nanofluid 11:111–117
29. Liao Y, Song J, Li E et al (2012) Rapid prototyping of three-dimensional microfluidic mixers in glass by femtosecond laser direct writing. Lab Chip 12:746–749
30. Hanada Y, Sugioka K, Ishikawa IS et al (2011) 3D microfluidic chips with integrated functional microelements fabricated by a femtosecond laser for studying the gliding mechanism of cyanobacteria. Lab Chip 11:2109–2115
31. Osellame R, Taccheo S, Marangoni M et al (2003) Femtosecond writing of active optical waveguides with astigmatically shaped beams. J Opt Soc Am B 20:1559–1567
32. Cheng Y, Sugioka K, Midorikawa K et al (2003) Control of the cross-sectional shape of a hollow microchannel embedded in photostructurable glass by use of a femtosecond laser. Opt Lett 28:55–57
33. Sugioka K, Cheng Y, Midorikawa K et al (2006) Femtosecond laser microprocessing with three-dimensionally isotropic spatial resolution using crossed-beam irradiation. Opt Lett 31:208–210
34. He F, Xu H, Cheng Y et al (2010) Fabrication of microfluidic channels with a circular cross section using spatiotemporally focused femtosecond laser pulses. Opt Lett 35:1106–1108
35. Charles GD, Michael G, Block E et al (2012) Squier intuitive analysis of space-time focusing with double-ABCD calculation. Opt Express 20:14244–14259
36. He F, Cheng Y, Lin J et al (2011) Independent control of aspect ratios in the axial and lateral cross sections of a focal spot for three-dimensional femtosecond laser micromachining. New J Phys 13:083014 (13)
37. Ams M, Marshall G, Spence D et al (2005) Slit beam shaping method for femtosecond laser direct-write fabrication of symmetric waveguides in bulk glasses. Opt Express 13:5676–5681
38. Sowa S, Watanabe W, Tamaki T et al (2006) Symmetric waveguides in poly (methyl methacrylate) fabricated by femtosecond laser pulses. Opt Express 14:291–297

Chapter 5
Fabrication of Fluid Control Microdevices

Abstract Fluid control microdevices such as microvalves, micropumps, and micromixers are key components in microfluidic systems. Femtosecond laser direct writing of glass followed by wet chemical etching (i.e., femtosecond-laser-assisted wet chemical etching) allows freely movable structures that are internally encapsulated inside microfluidic devices to be fabricated by a single continuous process. Such structures can function as fluid control microdevices that can be used to control the flow rate and direction of fluids in microfluidic channels. This chapter describes fabrication of a freely movable microplate and microrotor whose motions are controlled by air pressure, optical force, or an external micromotor. The microplate functions as a microvalve, whereas the microrotor functions as a micropump.

5.1 Introduction

Microfluidic systems are a critical component in biomicrochips such as microfluidic, optofluidic, and lab-on-a-chip devices and micro total analysis systems (μ-TAS) for field-deployable chemical/biological analysis and medical examination with high efficiency, high accuracy, and high performance. They basically consist of microfluidic components such as microreservoirs and microfluidic channels for infusing, reacting, and storing liquid samples. To introduce liquid samples into microfluidic channels, a micropump is often externally connected to the microfluidic chips. To miniaturize microchips, it is desirable to internally integrate micropumps. Further, to control fluid flow and reactions in microfluidic systems, some other fluid control microdevices such as microvalves and micromixers should be also integrated. As described in Chap. 4, three-dimensional (3D) microreservoirs and microfluidic channels can be directly formed inside glass by femtosecond laser direct writing followed by wet chemical etching (i.e., femtosecond-laser-assisted wet chemical etching) [1–10]. This technique can be extended to fabricate freely movable microcomponents internally

K. Sugioka and Y. Cheng, *Femtosecond Laser 3D Micromachining for Microfluidic and Optofluidic Applications*, SpringerBriefs in Applied Sciences and Technology, DOI: 10.1007/978-1-4471-5541-6_5, © The Author(s) 2014

encapsulated inside microfluidic systems, so-called "ship-in-a-bottle" structures, which function as fluid control microdevices [11–14].

This chapter describes fabrication of "ship-in-a-bottle" structures inside glass by femtosecond-laser-assisted wet chemical etching to realize internal integration of microvalves and micropumps. The fabricated microstructures can be driven by air pressure, optical force, or an external micromotor to control the fluid flow in microfluidic systems.

5.2 Microvalve Fabrication

A freely movable glass microplate has been fabricated inside photosensitive glass by femtosecond-laser-assisted wet chemical etching using the same fabrication procedure as that described in Chap. 4 for fabricating microfluidic systems [11]. Figure 5.1 shows the exposure scheme for photosensitive glass used to fabricate this structure. After scanning the femtosecond laser beam in the dark regions in Fig. 5.1a, post-annealing was performed to form modified regions consisting of lithium metasilicate crystallites. These modified regions were then completely removed by wet chemical etching in dilute hydrofluoric acid and hollow structures were formed in the glass, as indicated by the white regions in Fig. 5.1b. Finally, a freely movable glass plate remained in a hollow structure.

The fabricated movable microplate functioned as a microvalve in microfluidic structures, as shown in Fig. 5.2. This device was manufactured by stacking three photosensitive glass substrates (each substrate was 2 mm thick since this was the thickest substrate available). A syringe supplied compressed air to move the plate.

Fig. 5.1 Schematic diagram of femtosecond laser exposure for fabricating a freely movable microplate embedded in photosensitive glass. **a** The focused femtosecond laser beam was scanned in the dark regions. **b** The dark regions were completely removed by post-annealing and wet chemical etching in dilute hydrofluoric acid. A movable glass plate remained in the hollow structure (white region) in the glass [11]

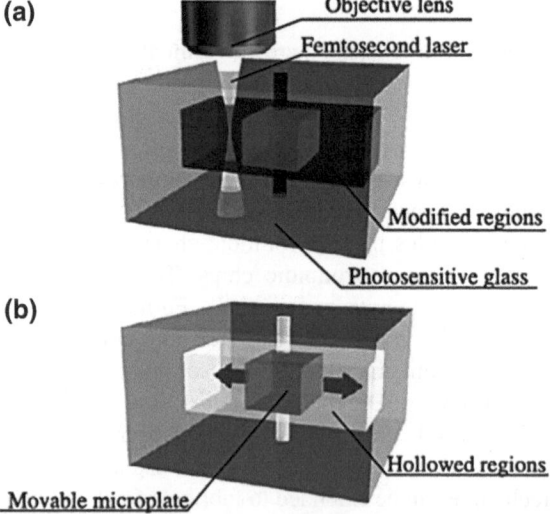

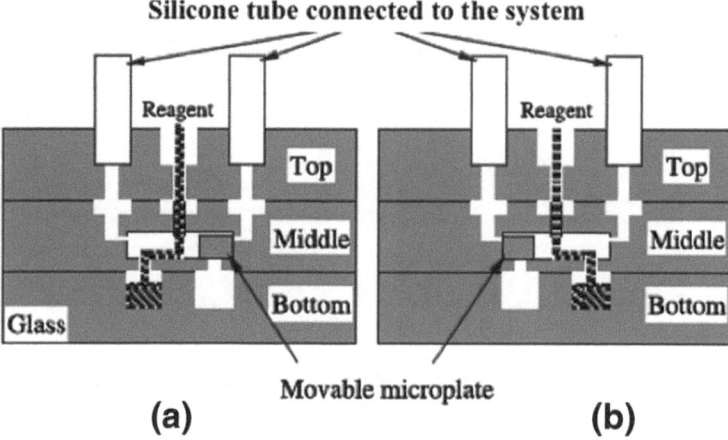

Fig. 5.2 Schematic diagram showing microplate functioning as a microvalve in a microfluidic system. **a** When compressed air is supplied from the left opening in the top layer, the microplate moves to the right. Then, the flow channel of the liquid sample to the right outlet is shut off so that the liquid sample only flows to the left outlet. **b** When compressed air is supplied from the right opening, the microplate moves to the left. Consequently, the fluid flows to the right outlet [11]

The top layer contained three inlet microreservoirs, one with an opening for infusing a liquid sample (central microreservoir) and two with openings attached to silicone tubes for supplying compressed air from the syringes (left and right microreservoirs). In the middle layer, a movable microplate was embedded in a rectangular hollow chamber connected by five microchannels to the microreservoirs in the top and bottom plates; the microplate was fabricated using the exposure scheme shown in Fig. 5.1. In the bottom layer, two microreservoirs were installed in the glass as outlets for the liquid samples.

When compressed air was supplied from the left opening in the top layer, the microplate moved to the right. In this case, the flow channel of the liquid sample to the right outlet was shut off so that the liquid sample only flowed to the left outlet (Fig. 5.2a). When compressed air was supplied from the right opening, the microplate moved to the left. Consequently, the fluid flowed to the right outlet (Fig. 5.2b). Thus, this microplate can switch the flow direction of liquids and it can thus function as a microvalve.

Figure 5.3 shows photographs of a fabricated sample. They show that the microplate moves laterally when the direction of the compressed air flow is switched. A liquid sample was introduced into the fabricated microfluidic system and it was confirmed that the microplate could switch the flow direction of the fluid.

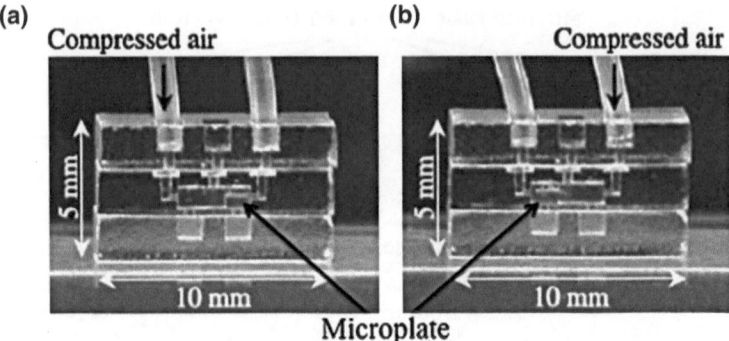

Fig. 5.3 Photographs of a fabricated microfluidic system in which a freely movable microplate is embedded. **a** When compressed air is supplied through the left opening in the top layer, the microplate moves to the right. **b** When compressed air is supplied through the right opening, the microplate moves to the left [11]

5.3 Microrotor Fabrication

A microrotor was fabricated in a cavity inside silica by femtosecond-laser-assisted wet chemical etching [12, 13]. Figure 5.4a shows a schematic diagram and Fig. 5.4b shows a phase contrast image of the fabricated microrotor. It has four blades and it is four-fold rotationally symmetric with respect to the rotation axis. It can be rotated by laser trapping induced by a focused beam incident on the side surfaces of the rotator [15]. Light refracted from these side surfaces generates radiation pressure, which produces a rotational torque in the four-blade rotor because the forces on all the side surfaces are directed in the same direction. Figure 5.5 shows sequential optical microscopy images of the rotating microrotator being rotated by laser trapping. The rotor rotated counterclockwise. The rotational speed can be controlled by varying the laser trapping power. It is nearly proportional to the laser power between 200 mW and 4 W when a 40 × objective lens was used for trapping, as shown in Fig. 5.6 (no rotation occurs at powers

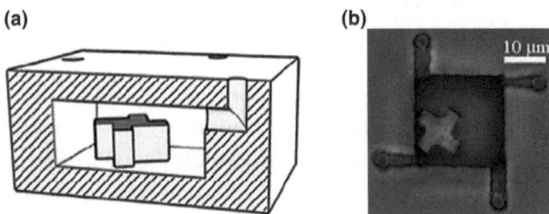

Fig. 5.4 **a** Schematic cross-sectional diagram and **b** phase-contrast image of optical microrotor encapsulated in silica. The microrotator is about 10 μm high by 10 μm wide and it is rotated using laser trapping [13] (Reproduced with permission from JLPS. ©2009 by Japan Laser Processing Society)

Fig. 5.5 Sequential optical microscope images (30 s^{-1}) of microrotator being rotated by laser trapping induced by a 4-W laser beam focused by a 100 × objective lens. The white arrow indicates the rotation direction [13] (Reproduced with permission from JLPS. ©2009 by Japan Laser Processing Society)

below 200 mW). The smaller numerical aperture (NA) of a 10 × objective lens produced a smaller rotation speed. Microrotors with different shapes such as a sloping top rotor can easily be fabricated by femtosecond-laser-assisted wet chemical etching, allowing a larger rotation speed to be generated with a low NA lens [13].

A similar microrotor was fabricated in a microfluidic system made of photo-sensitive glass. It functions as a micropump to control the fluid flow rate [14]. Fluid flow was characterized in the microfluidic channel produced by fabricated microrotors with three different blade structures (four rectangular blades (Fig. 5.7a), four triangular blades (Fig. 5.7b), and eight triangular blades (Fig. 5.7c)). The microrotors were rotated by an external DC micromotor. Figure 5.8 shows the dependence of the water flow rate in the microfluidic channel on the rotation speed of the microrotor for each shape. The flow rate increases linearly with increasing rotation speed. The microrotor with eight blades generated a higher flow rate than those with four blades. While the flow rate is almost independent of the blade shape, triangular blades generate less pulsation. Thus, the flow rate can be controlled over a wide range by controlling the rotation speed and the number of blades. Microfluidic components integrated with a micropump have been used to observe the rheotaxis behavior of aquatic microorganisms [14].

Fig. 5.6 Rotational speed of two different optical rotors as a function of laser trapping power [13] (Reproduced with permission from JLPS. ©2009 by Japan Laser Processing Society)

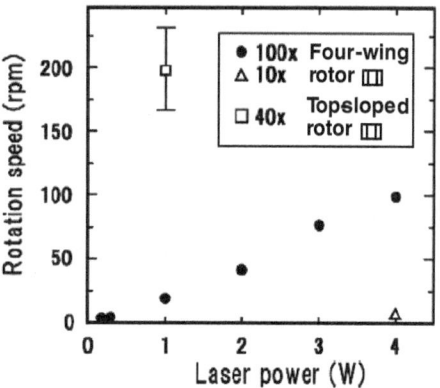

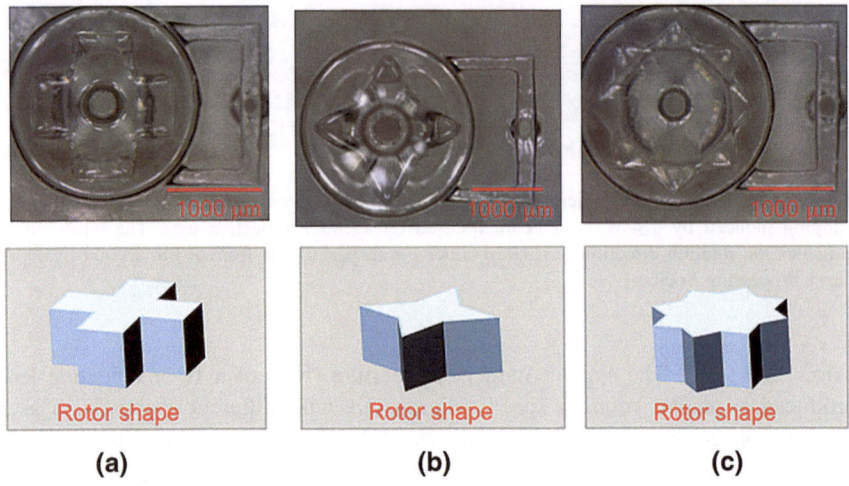

Fig. 5.7 Microrotors with three different blade structures fabricated in photosensitive glass: **a** four rectangular blades, **b** four triangular blades, and **c** eight triangular blades [14]

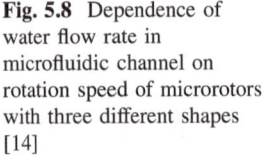

Fig. 5.8 Dependence of water flow rate in microfluidic channel on rotation speed of microrotors with three different shapes [14]

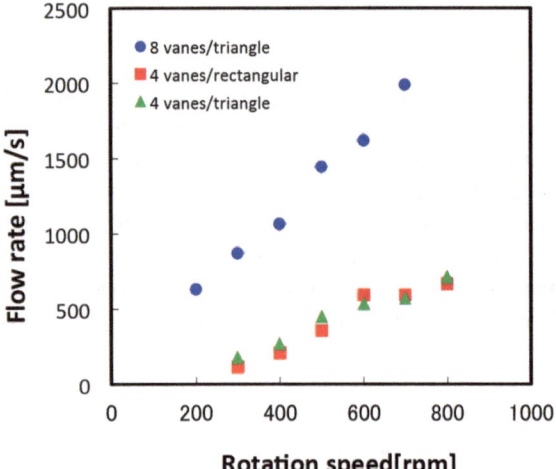

5.4 Summary

Femtosecond-laser-assisted wet chemical etching can fabricate not only 3D microfluidic structures but also freely movable structures encapsulated in microfluidic structures (i.e., "ship-in-a-bottle" structures). Using this technique, freely movable microplates and microrotors were embedded in microfluidic systems; they respectively function as microvalves for switching the fluid flow direction and as a micropump for controlling the flow rate. This technique has the advantage that a single continuous process can simultaneously fabricate microfluidic components and movable microcomponents with almost arbitrary 3D shapes without using complex stacking and bonding procedures. It is promising for integrating fluid control microdevices such as microvalves, micropumps, and micromixers in microfluidic systems.

References

1. Marcinkevicius A, Juodkazis S, Watanabe M et al (2001) Femtosecond laser-assisted three-dimensional microfabrication in silica. Opt Lett 26:277–279
2. Masuda M, Sugioka K, Cheng Y et al (2003) 3-D microstructuring inside photosensitive glass by femtosecond laser excitation. Appl Phys A 76:857–860
3. Sugioka K, Cheng Y, Midorikawa K (2005) Three-dimensional micromachining of glass using femtosecond laser for lab-on-a-chip device manufacture. Appl Phys A 81:1–10
4. Hanada Y, Sugioka K, Kawano H et al (2008) Nano-aquarium for dynamic observation of living cells fabricated by femtosecond laser direct writing of photostructurable glass. Biomed Microdevices 10:403–410
5. Hanada Y, Sugioka K, S-Ishikawa I et al (2008) 3D microfluidic chips with integrated functional microelements fabricated by a femtosecond laser for studying the gliding mechanism of cyanobacteria. Lab Chip 11:2109–2115
6. Bellouard Y, Said A, Dugan M et al (2004) Fabrication of high-aspect ratio, micro-fluidic channels and tunnels using femtosecond laser pulses and chemical etching. Opt Express 12:2120–2129
7. Kiyama S, Matsuo S, Hashimoto S et al (2009) Examination of etching agent and etching mechanism on femtosecond laser microfabrication of channels inside vitreous silica substrates. J Phys Chem C 113:11560–11566
8. Hnatovsky C, Taylor RS, Simova E et al (2005) Polarization-selective etching in femtosecondlaser-assisted microfluidic channel fabrication in fused silica. Opt Lett 30:1867–1869
9. Crespi A, Gu Y, Ngamsom B et al (2010) Three-dimensional Mach-Zehnder interferometer in a microfluidic chip for spatially-resolved label-free detection. Lab Chip 10:1167–1173
10. Schaap A, Rohrlack T, Bellouard Y (2012) Optical classification of algae species with a glass lab-on-a-chip. Lab Chip 12:1527–1532
11. Masuda M, Sugioka K, Cheng Y et al (2004) Direct fabrication of freely movable microplate inside photosensitive glass by femtosecond laser for lab-on-chip application. Appl Phys A 78:1029–1032
12. Matsuo S, Kiyama S, Shichijo Y et al (2008) Laser microfabrication and rotation of ship-in-a-bottle optical rotators. Appl Phys Lett 93:051107
13. Kiyama S, Tomita T, Matsuo S et al (2009) Laser fabrication and manipulation of an optical rotator embedded inside a transparent solid material. J Laser Micro/Nanoengin 4:18–21
14. Hanada Y, Iida M, Sugioka K (2009) Micro pump fabrication by femtosecond laser direct writing for microorganism analysis in nano-aquarium. CLEO/Pacific rim extended abstract:WJ1-4
15. Higurashi E, Ukita H, Tanaka H et al (1994) Optically induced rotation of anisotropic microobjects fabricated by surface micromachining. Appl Phys Lett 64:2209–2210

Chapter 6
Fabrication of Micro-optical Components in Glass

Abstract Fabrication of optofluidic systems requires synergetic incorporation of micro-optical components in microfluidic networks. This chapter describes key techniques for fabricating optical waveguides and free-space micro-optical components in glass by femtosecond laser microprocessing, both of which are essential building blocks for optofluidic devices. It is straightforward to fabricate optical waveguides as femtosecond laser irradiation can change the refractive index of glass through multiphoton absorption in the focal volume. Free-space micro-optical components such as micromirrors and microlenses can be fabricated using femtosecond-laser-assisted wet chemical etching to form hollow structures with planar or curved surfaces in glass that serve as optical interfaces. Micro-attenuators can be embedded in glass with a high spatial resolution and controllable attenuations by synthesizing silver nanoparticles in photosensitive glass using femtosecond laser irradiation and subsequent heat treatment.

6.1 Introduction

It is not easy to incorporate optical functions in microfluidic systems. The main difficulty is that most microfluidic systems are currently fabricated using lithography-based planar technologies, which lack the flexibility and/or capability to form three-dimensional (3D) micro-optical elements. Although some optical elements have been successfully incorporated in microfluidic chips by post-assembling, this increases the fabrication complexity and cost [1, 2]. Ideally, a technology will be realized that can simultaneously fabricate 3D microfluidic and 3D micro-optical structures in a single substrate. Femtosecond laser microfabrication has been the only method that has demonstrated the compatibility and high flexibility to fulfill this requirement.

In this chapter, we discuss the technical details of fabricating several key optical components in glass using femtosecond laser direct writing, including optical waveguides (Sect. 6.2), micromirrors (Sect. 6.3), microlenses (Sect. 6.4), and

K. Sugioka and Y. Cheng, *Femtosecond Laser 3D Micromachining for Microfluidic*
and Optofluidic Applications, SpringerBriefs in Applied Sciences and Technology,
DOI: 10.1007/978-1-4471-5541-6_6, © The Author(s) 2014

micro-attenuators (Sect. 6.5). We focus on fabrication of micro-optical components in photosensitive Foturan glass and fused silica because a rich variety of 3D microfluidic devices/systems have already been demonstrated in these two glasses using femtosecond lasers, as has been described in Chaps. 4 and 5. Thus, constructing integrated optofluidic devices will be straightforward, as is discussed in Chaps. 8 and 9.

6.2 Optical Waveguides

Optical waveguides are essential building blocks of photonic circuits and devices such as optical ring cavities, power splitters, directional couplers, Mach–Zehnder interferometers, Bragg grating waveguides, and waveguide lasers. In the mid-1990s, Hirao's group pioneered the writing of optical waveguides in fused silica by inducing permanent refractive index changes in the focal volumes of tightly focused femtosecond laser pulses [3]. To date, femtosecond laser direct writing has been used to produce optical waveguides in various transparent materials, including glass [4, 5], crystalline materials [6, 7], and polymers [8]. Writing waveguides in fused silica and photosensitive glass is particularly attractive for optofluidic applications since selective chemical etching can be locally induced in these two materials by femtosecond laser irradiation, making it possible to realize both optical and fluidic functions with a single exposure. The mechanism responsible for increasing the refractive index of glass during femtosecond laser irradiation has not been fully determined and is still being investigated. The current consensus is that refractive index modification can most likely be attributed to a combination of color-center formation and densification of glass; however, a quantitative model has yet to be developed [9, 10].

A transverse writing scheme is typically used to inscribe optical waveguides in fused silica and Foturan glass [11–13]; in other words, the sample is translated perpendicular to the incident beam (see Fig. 6.1a). This enables waveguides of arbitrary lengths and geometries to be written and is thus suitable for optofluidic integration. For optofluidic applications, the key optical parameters of a waveguide are its optical loss and mode profile, both of which critically depend on irradiation parameters such as the sample translation speed, the pulse energy, the pulse duration, the repetition pulse rate, and the numerical aperture of the focusing lens. An enormous number of combinations of these parameters have been used to write optical waveguides in glass that have satisfactory performances. There are two key criteria for optimizing the parameters. First, the peak intensity of the focused femtosecond laser pulses should be just slightly higher than the optical breakdown threshold to enable multiphoton excitation in glass; much higher intensities will result in the formation of microvoids and even cracks, which strongly scatter light. Second, based on the repetition rate and the focal spot size of the laser pulses, the translation speed should be controlled so that a sufficient pulse overlap can be

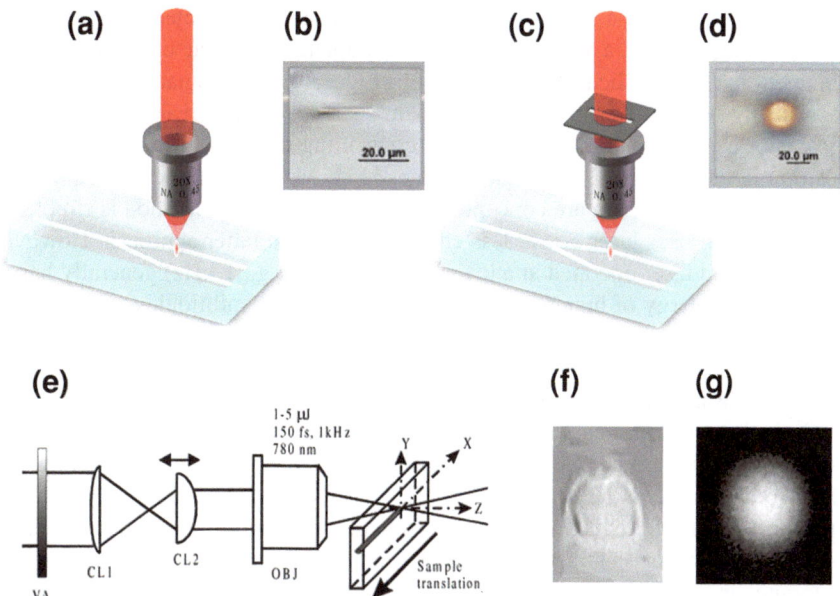

Fig. 6.1 **a** Schematic diagram of conventional transverse writing scheme, and **b** highly elliptical cross section of waveguide fabricated using focusing system in (**a**). **c** Schematic diagram of transverse writing scheme incorporating slit beam shaping, and **d** symmetric cross section of waveguide fabricated using focusing system in (**c**) [12]. (Reproduced with permission from OSA. ©2005 by the Optical Society of America). **e** Schematic diagram of transverse writing scheme incorporating astigmatic beam shaping technique. **f** Symmetric cross section, and **g** single-mode beam profile realized by waveguide fabricated using focusing system in (**e**) [11] (Reproduced with permission from OSA. ©2003 by the Optical Society of America)

realized to induce smooth refractive index modification while optimizing heat accumulation.

The mode profiles of the waveguides produced by the transverse writing scheme strongly reflect the intensity distribution in the plane parallel to the optical axis and perpendicular to the translation direction. Due to the focal spot being elongated in the propagation direction of the laser beam, waveguides produced by a transverse writing scheme intrinsically have elliptical cross sections, as shown by the waveguide inscribed in fused silica in Fig. 6.1b. Such an asymmetric cross section makes it difficult to achieve single mode propagation and thus is unsuitable for many applications [11, 12]. This problem can be resolved by intentionally expanding the focal spot in the transverse direction (i.e., the direction perpendicular to the laser beam propagation direction) using either astigmatic focusing or slit beam shaping [11–13]. The basic concept behind these shaping techniques is that by strongly compressing the incident beam in one direction before it enters the focal lens, tight focusing occurs only along the major axis of the elliptical incident beam. Thus, a disk-shaped focal spot is produced. A similar effect can also be achieved using deformable mirrors and adaptive slits [14–16]. Figure 6.1c shows a focusing system with a slit mounted directly in front of the objective lens. Using this focusing system, a waveguide with a completely circular cross section was obtained in fused silica (Fig. 6.1d) [12]. Figure 6.1e schematically depicts a focusing system employing astigmatic beam shaping, which creates a disk-shaped focal spot using an astigmatic beam and controlling the beam waist and the focal position in the tangential and sagittal planes [11]. Although slit beam shaping and astigmatic beam shaping are conceptually similar, the former method is simpler to implement but uses laser power less efficiently than the latter approach. However, a femtosecond laser operated at a low repetition rate (e.g., 1 kHz) generally has an output pulse energy of hundreds of microjoules or a few millijoules, which is two to three orders of magnitude higher than the pulse energy required for writing optical waveguides in glass. For such applications, energy loss due to the slit aperture will not be a problem.

Optical waveguides can also be inscribed in photosensitive Foturan glass using laser direct writing since laser irradiation increases the refractive index (Fig. 6.2) [17]. To obtain a circular cross section, which is required for a single-mode waveguide, a 0.2-mm-wide slit was inserted in front of the objective lens with a numerical aperture of 0.46 for shaping the focal spot during laser writing. Figure 6.2 shows near- and far-field beam profiles of two waveguides written at the same scan speed of 200 µm/s, but with different pulse energies of 1.8 and 2.25 µJ. It indicates that the waveguides exhibit single-mode propagation. Waveguides written in Foturan with higher pulse energies led to multi-mode beam profiles [17]. The optical loss of waveguides written under these conditions was measured to be ~ 0.5 dB/cm.

As mentioned above, both astigmatic focusing and slit beam shaping require that the samples be translated parallel to the major axis of the elliptical incident beam. Therefore, when fabricating curved waveguides, the orientation of the cylindrical telescope or the slit must be dynamically varied during laser writing [16]. To overcome this difficulty, an elegant technique has recently been developed to provide a 3D isotropic focal volume using spatio-temporally focused femtosecond laser pulses [18, 19]. Using this technique, microfluidic channels with circular cross sections have been fabricated for an arbitrary writing direction. The effectiveness of this technique for writing high-quality waveguides is currently being investigated. Chapter 4 describes beam shaping techniques for achieving circular cross sections in more detail.

Currently, waveguides written by femtosecond lasers in fused silica and Foturan typically have propagation losses of the order of 0.5–1 dB/cm, which are sufficiently low for most on-chip optofluidic applications.

6.3 Micro-optical Mirrors

As discussed in Chap. 4, hollow microfluidic structures can be formed in glass by femtosecond-laser-assisted wet chemical etching. This approach can be extended to fabricate 3D micro-optical components such as optical micromirrors,

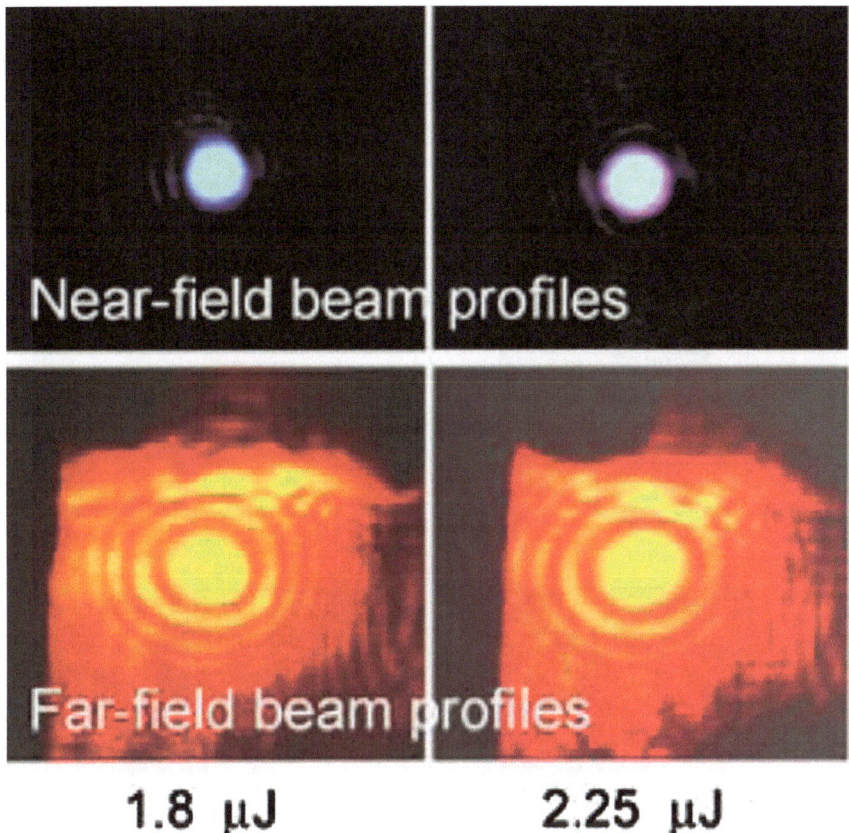

Fig. 6.2 Mode profiles of output from 10-mm-long optical waveguides fabricated using different laser energies and a scan speed of 200 µm/s [17]

microbeam splitters [20], and cylindrical and spherical microlenses in Foturan glass [21, 22], as well as spherical microlenses in fused silica that can realize nearly diffraction-limited focusing [23, 24].

Procedures for fabricating micro-optical components in Foturan glass mainly involve the following four steps: (1) direct writing of a latent image in the sample by a tightly focused femtosecond laser beam, (2) baking the sample in a programmable furnace to form modified regions, (3) etching the sample in 10 % hydrofluoric acid in an ultrasonic bath to selectively remove the modified regions, and (4) post-annealing the sample at 570 °C for 5 h to smooth the etched surface [20]. Step (4) is very important for fabricating high-performance optics, as discussed below.

Figure 6.3a shows a 45° micromirror that efficiently reflects a beam at right angles through total internal reflection. To form the micromirror, horizontal lines were scanned layer by layer from the top to the bottom surface of the glass sample

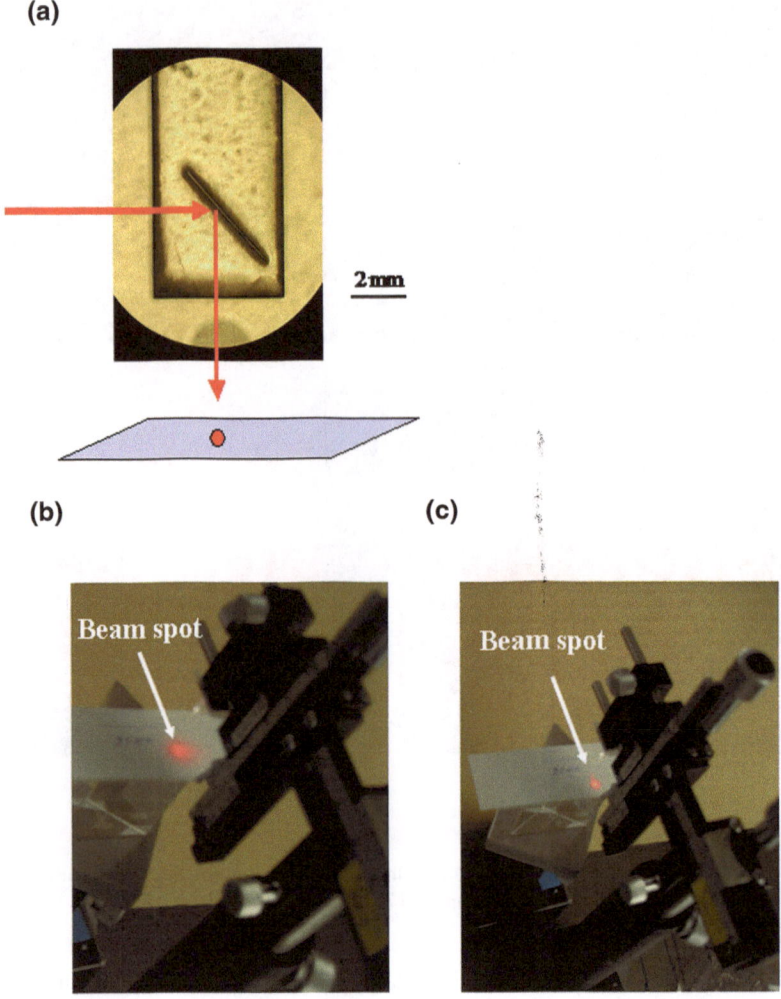

Fig. 6.3 a Top view of 3D micromirror fabricated inside Foturan glass (*arrows* indicate optical path) and beam spots reflected from sample, **b** before, and **c** after additional annealing

by tightly focused femtosecond laser pulses. After laser irradiation, the sample was subjected to heat treatment and chemical etching, resulting in the formation of a hollow planar structure vertically embedded in glass. The sample was then rinsed in distilled water and dried with nitrogen gas.

To examine the optical properties of the micromirror, the sidewalls of the etched sample were polished. A helium–neon (He–Ne) laser was used to examine the beam spot reflected from the etched inner surface, as shown in Fig. 6.3b. The reflected spot was recorded on a screen placed 10 mm from the end of the Foturan glass sample, as indicated by the arrow in Fig. 6.3b. The beam spot size was significantly larger than the initial spot size of the incident laser beam, implying that the reflected beam diverged strongly during propagation. Since lithium metasilicate crystallites produced by the heat treatment must be grown to a certain size (a few microns) to form an etchable network, etching of the crystallites left a rough surface, as shown in Fig. 6.4a. Such a high roughness will inevitably cause strong scattering and consequently divergence and loss of light.

Since the inner surface was not accessible for conventional mechanical polishing, the surface was smoothed using an additional annealing step. The temperature of this additional anneal was lower than that for crystallizing lithium metasilicate in Foturan glass. The temperature was ramped to 570 °C at 5 °C/min, held at this temperature for 5 h, and then reduced to 370 °C at 1 °C/min. The furnace was turned off and the sample was gradually cooled to room temperature. During this additional anneal, a thin liquid layer forms on the surface of the glass sample, significantly reducing the roughness through surface tension, as evidenced in Fig. 6.4b. This was found to reduce the average surface roughness from ∼81 nm to ∼0.8 nm. Consequently, the spot size of the reflected beam from the additionally annealed surface (as shown in Fig. 6.3c) appears much smaller than that in Fig. 6.3b, indicating a greatly reduced divergence angle and scattering loss.

The optical micromirrors have been used for constructing a compact optical circuit, as shown in Fig. 6.5a. Three 45° micromirrors were arranged in a small area of 4 mm × 5 mm. The circuit size was determined from the diameter of the He–Ne laser beam (∼1 mm) used in the experiment. After three consecutive reflections, the incident laser beam was rotated by 270° and it impinged on a screen placed 5 mm from the end of the sample, as shown in Fig. 6.5b. The reflected beam remained well collimated, as evidenced by the small beam spot on the screen.

The optical loss of the micromirror was examined at the communication wavelength of 1,550 nm. The micromirror exhibited an optical loss of only ∼0.24 dB, which is sufficiently low for many photonic applications. The micro-optical circuit was measured to have an optical loss of ∼1.6 dB. This higher loss is due to the slight curvature of the etched inner surface as a result of the relatively

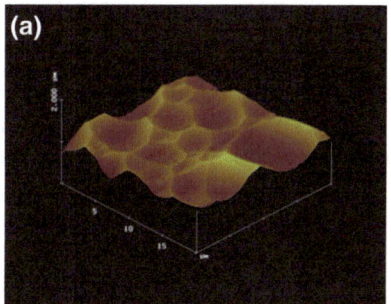
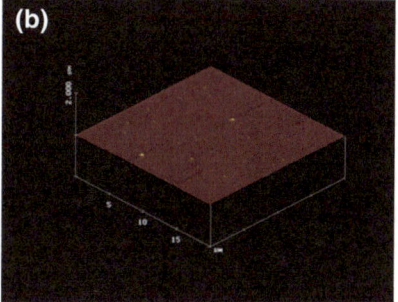

Fig. 6.4 AFM images of etched surfaces of Foturan glass after femtosecond laser irradiation and chemical etching **a** before, and **b** after additional annealing [20]

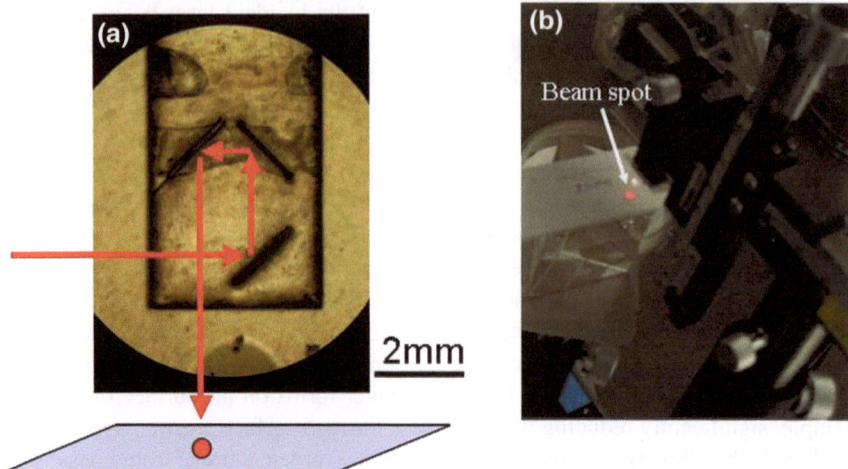

Fig. 6.5 Micro-optical circuit fabricated inside Foturan glass. **a** Top view of structure.
b Reflected beam spot on test screen

low contrast ratio of the selective chemical etching. This curvature distorts the
wavefront of the reflected beam. Consequently, after the relatively long optical
path (~ 14 mm) in the optical circuit, a portion of the light will leak from the
upper and bottom surfaces of the glass, increasing the optical loss.

6.4 Micro-optical Lenses

The same procedure for fabricating micromirrors can also be used to fabricate
micro-optical lenses in Foturan glass by forming curved inner surfaces rather than
planar surfaces [21, 22]. Figure 6.6a shows a micro-optical hemispherical lens
with a radius of 1 mm obtained using contour scanning. The focusing ability of the
microlens was examined using a He–Ne laser. The focal spot produced by the
microlens was projected onto a charge-coupled device (CCD) camera with a
10 × objective lens, as shown in Fig. 6.6b.

The latest progress in this area is the use of fused silica instead of Foturan. This
is advantageous since fused silica has superior optical properties to Foturan (e.g., a
wider transmission range and lower autofluorescence), which are highly desirable
for chip-based optical sensing applications [23]. The excellent homogeneity of
fused silica also results in the formation of smooth, regular surfaces over a large
area, leading to nearly diffraction-limited optical performance.

Figure 6.7 shows the process for fabricating micro-optical lenses in fused silica
[23, 24]. A layer-by-layer annular scanning method was used to inscribe the
contour of the microlens, as shown in Fig. 6.7a. To impart the hemispherical lens

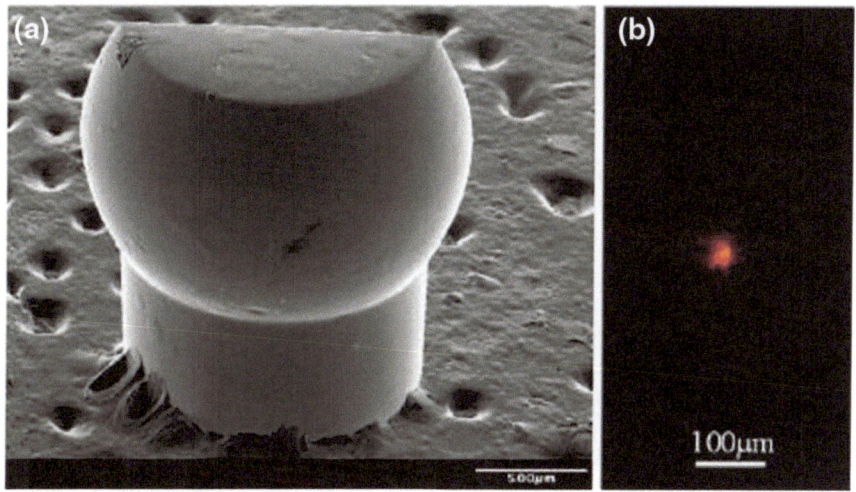

Fig. 6.6 **a** SEM image of a micro-optical hemispherical lens. **b** CCD camera image of a He–Ne laser beam focused by the micro-optical hemispherical lens in (**a**) [21]

with a high surface homogeneity, dynamic delta Z scanning was performed from bottom to top [25]. After laser irradiation, the sample was etched for ~80 min in an ultrasonic bath containing 10 % HF (Fig. 6.7b), until a microlens formed. The sample was then polished by surface reflow with an OH flame (JD90, Zhengzhou, China), as shown in Fig. 6.7c. By adjusting the flame flow, the temperature of the glass sample could be controlled to be close to the melting point of fused silica (1,730 °C). At such a high temperature, the glass surface melts slightly and forms a thin layer of a liquid-like phase. The surface tension of this thin liquid layer causes the microlens surface to become smooth.

Figure 6.8a and b shows the fabricated micro-optical lens before and after flame polishing, respectively. They show that the surface roughness is greatly reduced by the flame polishing. The measured focal spot in Fig. 6.8c has a similar diameter to that of the calculated focal spot, indicating that the microlens can focus a collimated beam to a nearly diffraction-limited focal spot. The microlens is thus capable of producing two-photon fluorescence images of biotissues with a quality comparable to that obtained by a commercially available objective lens [26].

6.5 Micro-optical Cavities

Micro-optical cavities with small sizes and ultra-high optical quality factors (Q factors) are critical on-chip optical devices for optical communication and biosensing [27, 28]. Fabrication of a microtoroidal cavity in fused silica has recently been demonstrated by femtosecond laser direct writing [29]. The fabrication process

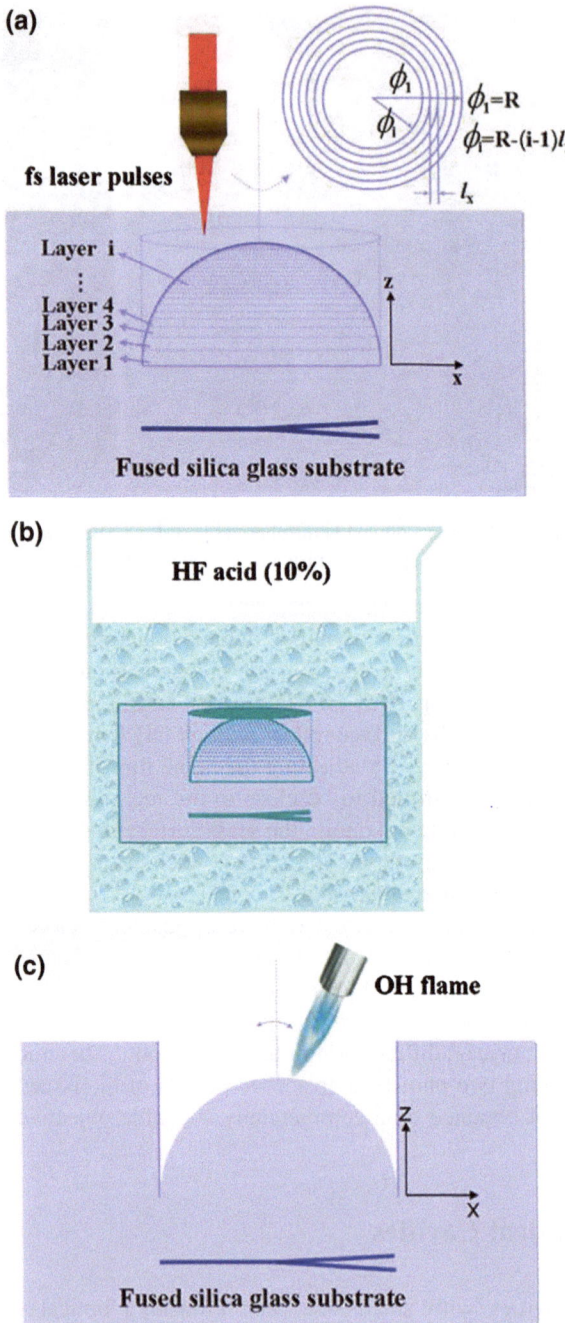

Fig. 6.7 Schematic showing process for fabricating microlens in fused silica [24]. **a** Layer-by-layer annular scanning process, **b** Chemical etching in HF acid solution, **c** Polishing process

(a)   **(b)**

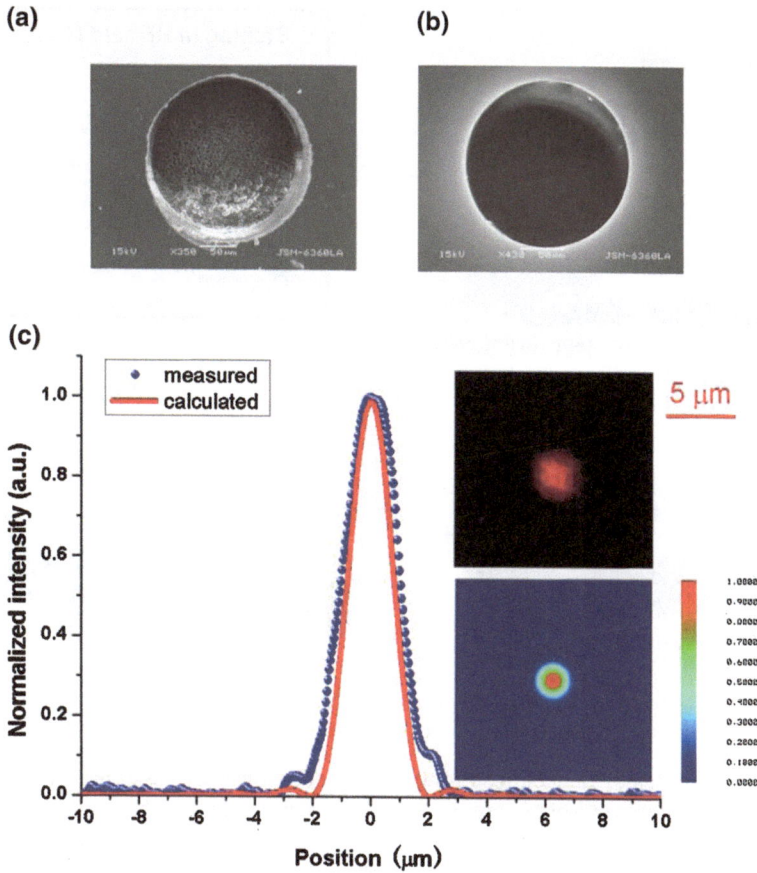

Fig. 6.8 Optical micrographs of fabricated micro-optical lens **a** before, **b** after flame polishing, and **c** focal spot produced by lens in (**b**). The *upper* and *lower* insets in (**c**) show the measured and calculated focal spot, respectively [23]

is schematically depicted in Fig. 6.9. It mainly consists of the following two steps: (1) femtosecond laser exposure followed by selective wet etching of the irradiated areas to create a microdisk structure; and (2) selective reflow of the silica cavities by CO_2 laser annealing to improve the quality factors. In femtosecond laser direct writing, a $100 \times$ objective with a numerical aperture (NA) of 0.9 was used to focus the beam to a ~ 1-μm-diameter spot. To form the microdisk supported by a thin pillar, a layer-by-layer annular scanning method was adopted by setting the lateral scanning step to be 1 μm.

Unlike microtoroidal cavities fabricated by conventional lithography which inherently have an in-plane geometry, the disk plane of the microcavity fabricated by femtosecond laser direct writing can be tilted at any angle with respect to the

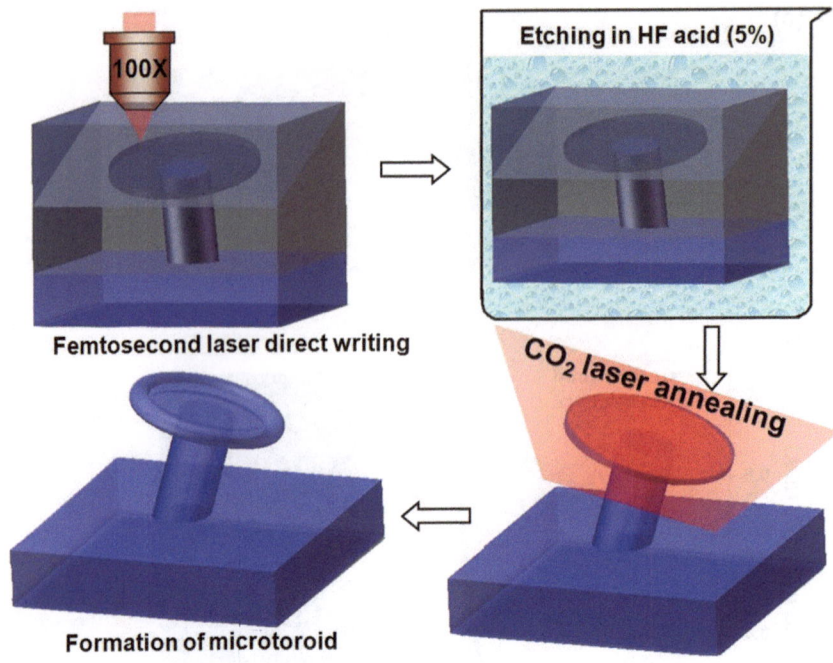

Fig. 6.9 Processes for fabricating a 3D microcavity by femtosecond laser direct writing

glass substrate, as evidenced by Fig. 6.10a. In addition, it is possible to form two closely located microtoroidal cavities with different heights on the same substrate, as shown in Fig. 6.10b. Fabrication of micro-optical cavities with out-of-plane geometries will provide flexibility for coupling light into and out of microcavities, which is promising for use in applications such as microlasers, sensors, and quantum information processors.

To characterize the mode structure and Q factor of the microtoroidal cavity, resonance spectra were measured by optical fiber taper coupling. Figure 6.11a shows a resonance transmission spectrum of a tapered fiber coupled to a microdisk cavity (diameter: ~ 38.3 μm; thickness: ~ 9 μm) with several excited whispering-gallery modes (WGMs). The free spectral range of 14.2 nm between two modes with successive angular mode numbers agrees well that calculated numerically based on the experimentally measured disk diameter. Figure 6.11b shows an individual WGM located at 1,562.85 nm with a Lorentzian shaped dip. The linewidth was measured to be 0.745 pm and the Q factor of the mode was calculated to be 2.1×10^6 based on the Lorentzian fit of the spectrum. This indicates that femtosecond laser micromachining of fused silica enables smooth cavity surfaces to be fabricated with low surface-scattering losses for WGMs.

Because of the high Q-factors of 3D microtoroidal cavities, their integration with microfluidic structures will enable biosensing with a sensitivity down to the single

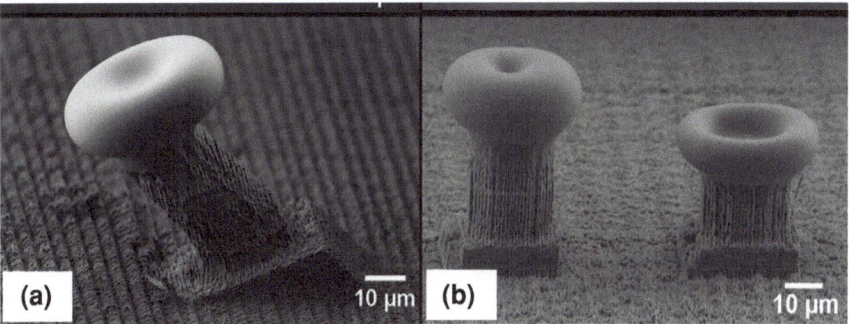

Fig. 6.10   SEM images of **a** a micro-optical cavity with a tilted disk plane, and **b** two microtoroidal cavities with different heights. Both structures are formed on fused silica substrates [29]

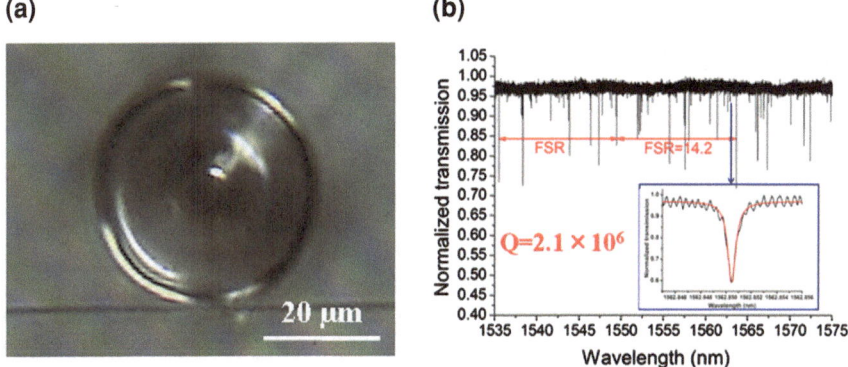

Fig. 6.11   a Transmission spectrum of microcavity coupled with tapered fiber. The free spectral range of 14.2 nm agrees well that calculated numerically. **b** Lorentzian fit (*red solid line*) of measured spectrum around the resonant wavelength of 1,562.85 nm (*black dotted line*), indicating a Q factor of 2.1 × 10^6 [29]

molecule level [28]. Femtosecond laser direct writing is a promising technique for such applications due to its ability to simultaneously produce microcavities and microfluidic structures such as microchannels and microchambers [30, 31].

6.6   Micro-optical Attenuators

Optical attenuators are important components for controlling the intensity of the incident light on samples being analyzed in microanalysis systems. For optofluidic applications, micro-optical attenuators can be directly fabricated in Ag$^+$-doped

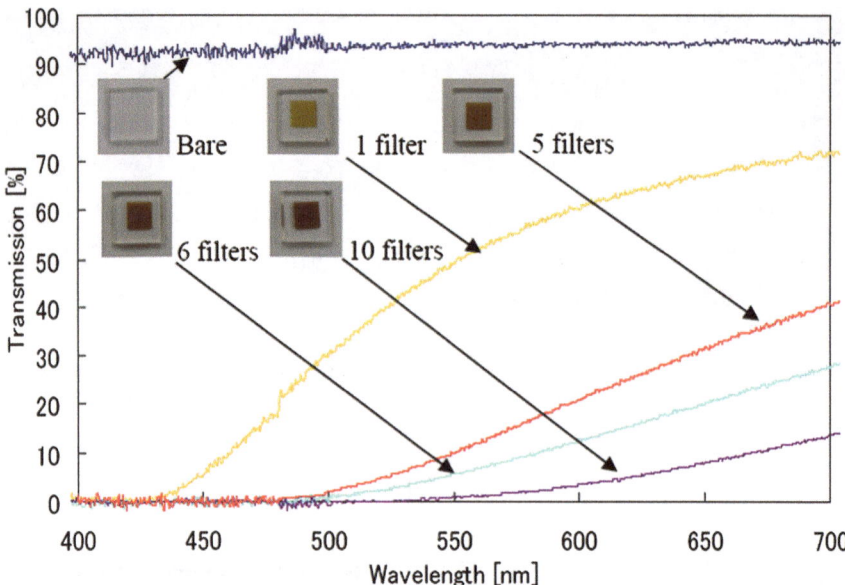

Fig. 6.12 Transmission spectra of untreated Foturan glass and Foturan glass containing one, five, six, and ten optical filter layers [32]

photosensitive Foturan glass by femtosecond laser irradiation followed by post-annealing as this process forms a layer of a brown crystalline phase of lithium metasilicate [32]. On femtosecond laser irradiation, nonlinear absorption generates free electrons in Foturan, which reduce silver ions doped in Foturan to silver atoms. By post-annealing the substrate, the laser-irradiated regions will turn brown due to the formation of lithium metasilicate crystallites (see Fig. 6.12), which act as optical attenuators.

The annealing conditions for fabricating optical attenuators are the same as those for fabricating microfluidic structures in Foturan glass (see Chap. 4). Briefly, a glass sample is annealed in a programmable furnace to form the crystalline phase of lithium metasilicate. The temperature was first increased to 500 °C at 5 °C/min, held constant for 1 h, raised to 605 °C at 3 °C/min, and then held constant for another 1 h. The attenuation of a single layer of the optical attenuator was constant for the same irradiation and annealing conditions. This permitted attenuation to be precisely controlled by stacking multiple layers of optical attenuators. To form a single optical attenuator layer, a focused laser beam was scanned line-by-line across the glass in the X–Y plane. Multiple layers were formed to reduce light transmission by shifting the laser focus position in the z-direction after forming preceding layers and then rescanning in the x–y plane.

The insets in Fig. 6.12 show photographs of optical attenuators with different attenuations, while Fig. 6.12 shows their corresponding transmittance curves in the visible region [32]. The attenuators have dimensions of 5 mm × 5 mm.

The average transmittance in the wavelength range 400–700 nm is evaluated to be 41, 14, 9, and 3 % for one, five, six, and ten layers. Attenuators with higher attenuations appear darker due to their greater absorption. These attenuators have been used to investigate the mechanism of micro-organism movement in a liquid environment. This movement is sensitive to the illumination intensity [32], as is described in Chap. 9.

6.7   Summary

Femtosecond laser direct writing has been used to fabricate free-space and waveguide optical components in glass. Waveguides are written by inducing a refractive index increase in glass by femtosecond laser irradiation; whereas free-space optics such as micromirrors, microlens, and high-Q optical cavities are fabricated by femtosecond-laser-assisted wet chemical etching and subsequent heat treatment (to smooth the surfaces to optical quality through surface reflow). Waveguides written by femtosecond laser irradiation in glass typically have propagation losses of the order of 0.5–1 dB/cm. On the other hand, micro-optical lenses fabricated by femtosecond-laser-assisted wet chemical etching can produce nearly diffraction-limited focal spots and realize high-quality two-photon fluorescence imaging of biotissues.

In photosensitive Foturan glass, selective precipitation of silver nanoparticles by femtosecond laser direct writing and subsequent growth of crystalline lithium metasilicate by post-annealing can be induced in the irradiated region, enabling micro-optical attenuators with arbitrary attenuations to be produced.

Waveguides and free-space optics can be monolithically integrated in microfluidic systems by femtosecond laser microprocessing, as discussed in Chaps. 8 and 9.

References

1. Tung YC, Zhang M, Lin CT et al (2004) PDMS-based opto-fluidic micro flow cytometer with two-color, multi-angle fluorescence detection capability using PIN photodiodes. Sensor Actuat B 98:356–367
2. Lien V, Vollmer F (2007) Microfluidic flow rate detection based on integrated optical fiber cantilever. Lab Chip 7:1352–1356
3. Davis KM, Miura K, Sugimoto N et al (1996) Writing waveguides in glass with a femtosecond laser. Opt Lett 21:1729–1731
4. Schaffer CB, Brodeur A, García JF et al (2001) Micromachining bulk glass by use of femtosecond laser pulses with nanojoule energy. Opt Lett 26:93–95
5. Li ZL, Low DKY, Ho MK et al (2006) Fabrication of waveguides in Foturan by femtosecond laser. J Laser Appl 18:320–324
6. Dong NN, Mendivil JM, Cantelar E et al (2011) Self-frequency-doubling of ultrafast laser inscribed neodymium doped yttrium aluminum borate waveguides. Appl Phys Lett 98: 81103(3)

7. Liao Y, Xu J, Cheng Y et al (2008) Electro-optic integration of embedded electrodes and waveguides in LiNbO$_3$ using a femtosecond laser. Opt Lett 33:2281–2283

8. Sowa S, Watanabe W, Tamaki T et al (2006) Symmetric waveguides in poly (methyl methacrylate) fabricated by femtosecond laser pulses. Opt Express 14:291–297

9. Reichman WJ, Krol DM, Shah L et al (2006) A spectroscopic comparison of femtosecond-laser-modified fused silica using kilohertz and megahertz laser systems. J Appl Phys 99:123112(5)

10. Ponader C, Schroeder J, Streltsov A (2008) Origin of the refractive-index increase in laser-written waveguides in glasses. J Appl Phys 103:063516(5)

11. Osellame R, Taccheo S, Marangoni M et al (2003) Femtosecond writing of active optical waveguides with astigmatically shaped beams. J Opt Soc Am B 20:1559–1567

12. Ams M, Marshall G, Spence D et al (2005) Slit beam shaping method for femtosecond laser direct-write fabrication of symmetric waveguides in bulk glasses. Opt Express 13:5676–5681

13. Cheng Y, Sugioka K, Midorikawa K et al (2003) Control of the cross-sectional shape of a hollow microchannel embedded in photostructurable glass by use of a femtosecond laser. Opt Lett 28:55–57

14. Thomson RR, Bockelt AS, Ramsay E et al (2008) Shaping ultrafast laser inscribed optical waveguides using a deformable mirror. Opt Express 16:12786–12793

15. Salter PS, Jesacher A, Spring JB et al (2012) Adaptive slit beam shaping for direct laser written waveguides. Opt Lett 37:470–472

16. Zhang Y, Cheng G, Huo G et al (2009) The fabrication of circular cross-section waveguide in two dimensions with a dynamical slit. Laser Phys 19:2236–2241

17. Wang Z, Sugioka K, Hanada Y et al (2007) Optical waveguide fabrication and integration with a micro-mirror inside photosensitive glass by femtosecond laser direct writing. Appl Phys A 88:699–704

18. He F, Xu H, Cheng Y et al (2010) Fabrication of microfluidic channels with a circular cross section using spatiotemporally focused femtosecond laser pulses. Opt Lett 35:1106–1108

19. Durfee CG, Greco M, Block E et al (2012) Intuitive analysis of space-time focusing with double-ABCD calculation. Opt Express 20:14244–14259

20. Cheng Y, Sugioka K, Midorikawa K et al (2003) Three-dimensional micro-optical components embedded in photosensitive glass by a femtosecond laser. Opt Lett 28:1144–1146

21. Cheng Y, Tsai HL, Sugioka K et al (2006) Fabrication of 3D microoptical lenses in photosensitive glass using femtosecond laser micromachining. Appl Phys A 85:11–14

22. Wang Z, Sugioka K, Midorikawa K (2007) Three-dimensional integration of microoptical components buried inside photosensitive glass by femtosecond laser direct writing. Appl Phys A 89:951–955

23. He F, Cheng Y, Qiao L et al (2010) Two photon fluorescence excitation with a microlens fabricated on the fused silica chip by femtosecond laser micromachining. Appl Phys Lett 96:041108(3)

24. Qiao LL, He F, Cheng Y et al (2011) A microfluidic chip integrated with a microoptical lens fabricated by femtosecond laser micromachining. Appl Phys A 102:179–183

25. Guo R, Xiao SZ, Zhai XM et al (2006) Micro lens fabrication by means of femtosecond two photon photopolymerization. Opt Express 14:810–816

26. Qiao L, He F, Wang C et al (2011) Fabrication of a micro-optical lens using femtosecond laser 3D micromachining for two-photon imaging of bio-tissues. Opt Commun 284:2988–2991

27. Armani DK, Kippenberg TI, Spillane SM et al (2003) Ultra-high-Q toroid microcavity on a chip. Nature 421:925–928

28. Armani AM, Kulkarni RP, Fraser SE et al (2007) Label-free, single-molecule detection with optical microcantilevers. Science 317:783–787

29. Lin JT, Yu SJ, Ma YG et al (2012) On-chip three-dimensional high-Q microcavities fabricated by femtosecond laser direct writing. Opt Express 20:10212–10217

30. Sugioka K, Cheng Y (2012) Femtosecond laser processing for optofluidic fabrication. Lab Chip 12:3576–3589
31. Osellame R, Hoekstra HJWM, Cerullo G et al (2011) Femtosecond laser microstructuring: an enabling tool for optofluidic lab-on-chips. Laser Photon Rev 5:442–463
32. Hanada Y, Sugioka K, Shihira-Ishikawa I et al (2011) 3D microfluidic chips with integrated functional microelements fabricated by a femtosecond laser for studying the gliding mechanism of cyanobacteria. Lab Chip 11:2109–2115

Chapter 7
Selective Metallization of Glass

Abstract Selective metallization of glass can be used to incorporate microelectronic components in microfluidic systems making it an important technique for further enhancing the functions of biochips. Both femtosecond-laser-assisted electroless plating and femtosecond laser surface modification combined with electroless plating can be used to selectively deposit thin metal films only on laser irradiated regions, even on the internal walls of microfluidic structures. Additionally, two-photon-induced metal ion reduction of a liquid or polymer containing metal ions by femtosecond laser direct writing can be used to fabricate three-dimensional metal microstructures on glass substrates that have a high electrical conductivity. These metallization techniques can be utilized to manufacture functional microcomponents including microheaters for space-selective control of temperature in microfluidic systems and surface-enhanced Raman scattering platforms for highly sensitive analysis of biochemical samples.

7.1 Introduction

As described in Chaps. 4–6, femtosecond laser processing can be used to fabricate not only microfluidic structures [1–3], but also fluid control microcomponents such as microvalves [4] and micropumps [5] and micro-optical components such as optical waveguides [6, 7], micromirrors [8], microlenses [9, 10], and optical filters [11], which is beneficial for fabricating highly functional biochips including optofluidics and micro-total analysis systems (μ-TAS). Incorporation of microelectronic components is further desired to enhance the functionalities of biochips. Such microelectronic components are realized by selective metallization of glass substrates. For example, micropatterned thin metal films act as microheaters to spatially control the temperature in a microfluidic system [12]. They can be also used to control the motion of fluid control components through electromagnetic force. Selective metallization for forming metal nanoparticles and metal nanostructures in microfluidic systems is also of great importance for highly sensitive

K. Sugioka and Y. Cheng, *Femtosecond Laser 3D Micromachining for Microfluidic and Optofluidic Applications*, SpringerBriefs in Applied Sciences and Technology, DOI: 10.1007/978-1-4471-5541-6_7, © The Author(s) 2014

analysis of samples in biochips based on surface-enhanced Raman scattering (SERS) [13].

Conventional techniques for depositing thin metal films on solid substrates employ vacuum evaporation [14], sputtering [15], or chemical vapor deposition [16]. However, these techniques require a resist process based on photolithography for micropatterning deposited thin metal films. Laser direct writing is a reliable alternative technique for directly patterning thin metal films without using photolithography [17]. For example, laser-induced plasma-assisted ablation (LIPAA) [18, 19] and laser-induced forward transfer (LIFT) [20, 21] using a nanosecond laser are attractive techniques for selective metallization of transparent materials. However, they can treat only the surfaces of glass substrates so that they cannot metallize the interiors of microfluidic systems. On the other hand, femtosecond lasers enable thin metal films to be formed on the internal walls of microfluidic systems in glass due to multiphoton absorption [6, 22], which is desirable for fabricating highly functional biochips.

This chapter reviews techniques for selectively metallizing the surfaces and interiors of glass substrates by femtosecond laser irradiation. It also introduces fabrication of three-dimensional (3D) metal microstructures based on femtosecond-laser-induced photoreduction.

7.2 Femtosecond-Laser-Assisted Electroless Plating

Femtosecond lasers have been used to generate free electrons at glass surfaces, which reduce metal ions in an electroless plating solution to selectively deposit thin metal films on laser-irradiated regions [23]. Figure 7.1 shows a schematic diagram of the experimental arrangement for selectively metallizing glass by femtosecond-laser-assisted electroless plating. The substrate used in this study is photosensitive

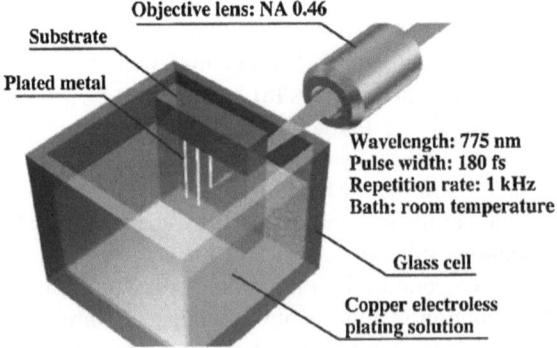

Fig. 7.1 Schematic diagram of experimental arrangement for selectively metallizing glass by femtosecond-laser-assisted electroless plating [23]

glass, which is commercially available under the trade name of Foturan. This glass is typically used for fabricating microfluidic systems by femtosecond laser irradiation (see Chap. 4). The glass substrate was placed in a fused silica cell filled with a copper electroless plating solution (no reducing agent was mixed in this case). A femtosecond laser beam (wavelength: 775 nm; pulse width: 140 ± 5 fs; repetition rate: 1 kHz) was focused on the rear surface of a glass substrate using an objective lens with a numerical aperture (NA) of 0.46 at room temperature. The laser spot size was estimated to be ∼20 μm in diameter. The cell in which the substrate was set was moved by a computer-controlled X–Y–Z stage during laser irradiation for metal patterning.

Figure 7.2 shows an optical microscopy image of the sample treated at a laser power of 4 mW and a scanning speed of 10 μm/s. A straight line of a copper thin film was selectively deposited on the photosensitive glass along the laser scanning direction. The linewidth of the deposited film was measured to be 20 μm, which is equal to the laser spot size. The as-deposited copper film could be easily peeled off, indicating that its adhesion was too low for practical applications. To improve the adhesion, the sample was annealed at 400 °C for 1 h after laser irradiation. The annealed sample passed the Scotch tape test. While the as-deposited copper film was less than 100 nm thick, self-aligned deposition of thicker films on the as-deposited regions can be performed by increasing the temperature of the plating solution to approximately 60 °C after laser irradiation. The deposited films had a comparable electrical conductivity to films prepared by conventional electroless plating.

One advantage of this technique is that it can selectively metallize internal walls of microfluidic structures embedded in glass since, by adjusting the laser energy, a femtosecond laser can induce multiphoton absorption at only the focal position even inside glass. To demonstrate this, a hollow microfluidic channel was fabricated inside photosensitive glass by femtosecond laser direct writing followed by thermal treatment and successive wet chemical etching in dilute hydrofluoric acid [2, 4]. The fabricated sample was immersed in the electroless copper plating solution and then irradiated by the femtosecond laser beam. The irradiation conditions were the same as those shown in Fig. 7.2. In this case, the laser beam was

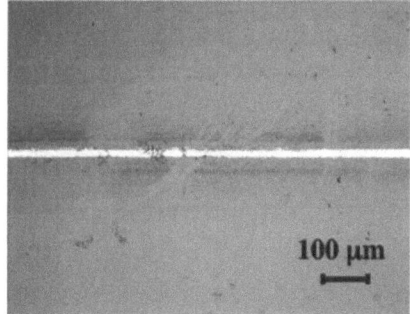

Fig. 7.2 Optical microscopy image of copper thin film selectively deposited on photosensitive glass by femtosecond-laser-assisted electroless plating at a laser power of 4 mW and a scanning speed of 10 μm/s [23]

100 μm

Fig. 7.3 Possible
mechanism for selective
metallization of glass by
femtosecond-laser-assisted
electroless plating [23]

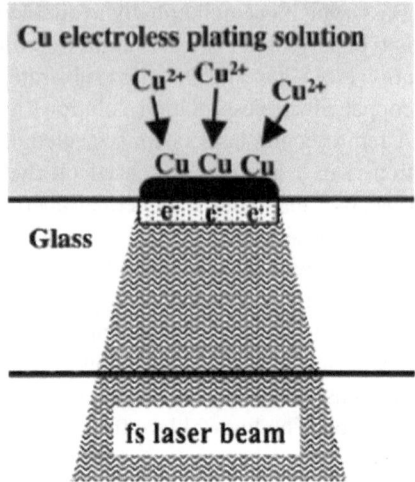

focused on one face of an internal wall in a microfluidic channel. A thin copper
film was successfully deposited only on the laser-irradiated regions of the internal
wall of the microfluidic channel inside the glass.

Figure 7.3 shows a possible mechanism for the selective metallization induced
by this technique. When a femtosecond laser beam is focused on the rear surface of
the photosensitive glass, multiphoton absorption occurs only at the focused region,
generating free electrons in the conduction band. The generated free electrons
reduce metal ions in the electroless plating solution near the laser-irradiated
regions, resulting in the precipitation of metal atoms and eventually the formation
of thin metal films on the laser-irradiated regions on the rear surface of the photosensitive glass. According to this mechanism, it should be possible to use this
technique to selectively metallize other glasses including fused silica.

7.3 Femtosecond Laser Modification Combined
with Electroless Plating

Another scheme that can selectively metallize glass is femtosecond laser modification combined with electroless plating. Although this technique involves multistep processes, it produces thin metal films that have a much higher adhesion than
those produced by the single-step process of femtosecond-laser-assisted electroless
plating described in Sect. 7.2.

7.3.1 Modification Using Silver Nitrate

Silver nitrate was used to form the seed layer for metallization by successive electroless plating [24]. Figure 7.4 illustrates the procedure for selective metallizing glass using silver nitrate ($AgNO_3$). It consists of four main steps: (1) Formation of $AgNO_3$ thin films on glass substrates; (2) modification of glass surfaces by femtosecond laser direct writing; (3) removal of unirradiated $AgNO_3$ films by acetone; and (4) selective copper coating by electroless plating. As a photosensitive layer, $AgNO_3$ films were prepared by immersing substrates into a 0.5 mol/l $AgNO_3$ solution and withdrawing the substrates at a speed of approximately 1 mm/s. The substrates were then naturally dried in the dark at room temperature for 12 h. The $AgNO_3$ film thickness could be controlled by adjusting the withdrawal rate. When a femtosecond laser beam is focused on glass surfaces coated with silver nitrate films, silver particles form due to decomposition of the silver nitrate film in the irradiated area. Femtosecond laser direct writing has a high spatial resolution and does not produce a significant heat affected zone, which facilitates the selective deposition of silver particles in the irradiated area. In the subsequent electroless plating, these particles can serve as seeds for selective copper deposition. Prior to electroless chemical plating, the substrates were cleaned with acetone in an ultrasonic bath to remove unirradiated $AgNO_3$ film. Finally, thin metal films were selectively deposited on laser-irradiated regions by electroless plating in a solution containing copper (II) sulfate pentahydrate ($CuSO_4 \cdot 5H_2O$), ethylenediaminetetraacetic acid (EDTA) disodium salt, and formaldehyde (HCHO) with concentrations of 5, 14, and 5 g/l, respectively (4 in Fig. 7.4).

To examine the electrical properties of copper microstructures, microelectric circuits were fabricated on glass substrates, as shown in Fig. 7.5a. Pattern I (Fig. 7.5b) and pattern II (Fig. 7.5c) had resistances of approximately 15 and 6 Ω, respectively. These results confirm that the copper microstructures formed on the glass substrates had high electrical conductivities. In addition, these selectively

Fig. 7.4 Schematic diagram showing process for selectively metallizing insulators: *1* formation of silver nitrate thin film on insulating substrate; *2* modification of insulator surface by femtosecond laser direct writing; *3* removal of unirradiated $AgNO_3$ film by acetone; and *4* copper coating by selective electroless plating [24]

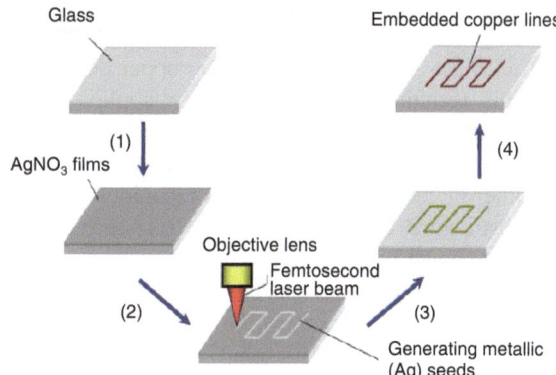

Fig. 7.5 Photographs of
metal patterning of glass
using a laser pulse energy of
8 μJ: **a** Electrical circuits
fabricated on glass surface;
b Cu lines in pattern I; and
c Cu lines in pattern II [24]

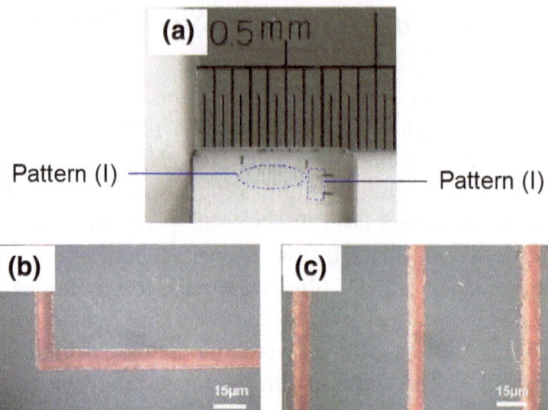

deposited copper lines exhibited good adhesion to the glass substrate, as they
remained intact after cleaning with distilled water in an ultrasonic bath. Their good
adhesion was further confirmed by a peel test using Scotch tape.

The mechanism for selective metallization of glass surfaces by femtosecond
laser irradiation was investigated by X-ray photoelectron spectroscopy (XPS) and
energy-dispersive X-ray (EDX) analysis [25]. For this analysis, a glass substrate
coated with $AgNO_3$ film was modified by femtosecond laser irradiation and then
ultrasonically cleaned in distilled water to remove unirradiated $AgNO_3$ film. When
the femtosecond laser beam is focused on the surface, silver particles form through
the decomposition of $AgNO_3$ films in the irradiated area [26, 27]. The decomposition reaction of silver nitrate is described by:

$$2AgNO_3 \rightarrow 2Ag + 2NO_2 \uparrow + O_2 \uparrow \tag{7.1}$$

After laser irradiation, although the sample was ultrasonically cleaned in distilled water for 10 min to remove unirradiated $AgNO_3$ film, selective electroless
copper plating can still be performed since femtosecond laser ablation induces
localized melting and quasi-welding at interfaces between the silver particles and
the glass surface [28, 29], which promotes good adhesion between the reduced
silver particles and the substrate. In the subsequent electroless plating, these
particles act as seeds for selective copper deposition.

This technique was successfully applied to demonstrate a Mach–Zehnder
interferometer electro-optic modulator [30] (see Chap. 8). It can form metal
microstructures on other insulators besides glass including crystals and polymers.
Another interesting feature of this technique is that it can deeply embed metals in
microstructures formed on the substrate surface through femtosecond laser ablation, as shown in Fig. 7.6.

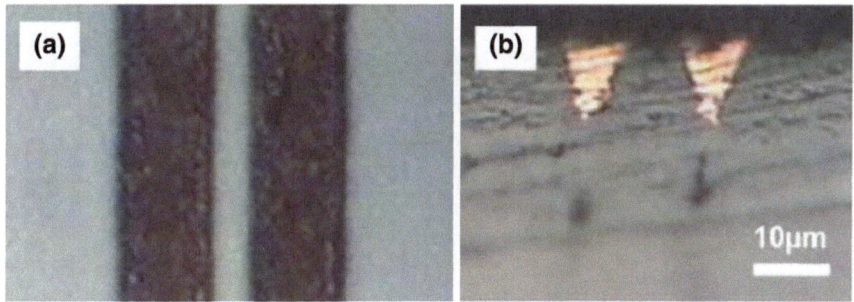

Fig. 7.6 Optical micrographs of microelectrodes embedded in LiNbO$_3$ crystal: **a** Top and **b** end views [24]

7.3.2 Modification by Ablation

Femtosecond laser ablation can produce seeds for selective metallization by successive electroless plating [12]. A femtosecond laser beam (wavelength: 775 nm; pulse width: 140 ± 5 fs; repetition rate: 1 kHz) was focused on the photosensitive glass surface using an objective lens with an NA of 0.46 to ablate the glass surface in air and the glass substrate was scanned using a PC-controlled X–Y–Z stage for micropatterning. The laser beam had a spot size of ∼5 μm. After ablation, the substrate was cleaned by ethanol and then by deionized water for 5 min each. Electroless copper plating was then performed at 50 °C for 20 min using a mixture of copper-ion solution and a reducing agent. This two-step process is performed to selectively deposit copper micropatterns on the laser-ablated regions.

As an application of this selective metallization technique, fabrication of a microheater was demonstrated. For this purpose, the focused femtosecond laser beam was first scanned on the glass surface in a continuous rectangular wave pattern. Electroless copper plating was then performed on the ablated glass sample. Figure 7.7a shows a microscope image of the fabricated microheater and Fig. 7.7b shows an enlarged image of the left side of the microheater. After fabricating the microheater, an electric power supply was connected to both ends using a silver adhesive. To measure the temperature increase due to the microheater, a laser thermometer was focused on the glass surface on which the microheater was fabricated. Figure 7.8 shows the relation between the temperature induced by the microheater and the heating power. The temperature increases almost linearly up to 200 °C with increasing heating power, which implies that the temperature can be easily controlled electrically. A temperature increase of 200 °C is sufficiently high for biochip applications. Furthermore, it was confirmed that the temperature could be reproducibly controlled.

This technique was used to selectively deposit a thin metal film on the internal wall of a microfluidic structure embedded in photosensitive glass. A 3D microfluidic channel connected to two microreservoirs (see Fig. 7.9a) was fabricated inside photosensitive glass by femtosecond laser direct writing followed by

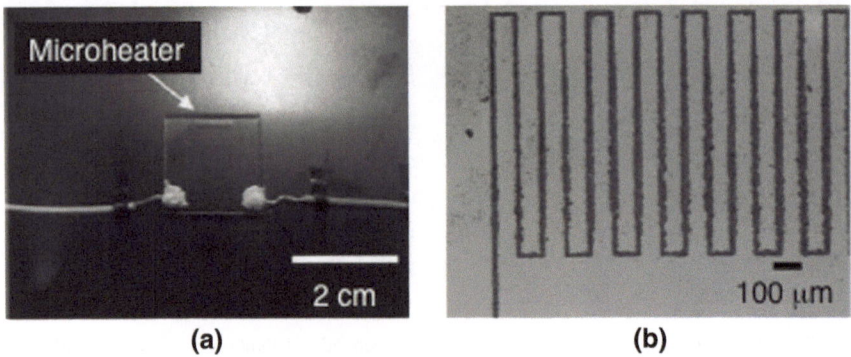

Fig. 7.7 **a** Microscope image of fabricated microheater and **b** enlarged image of left side of microheater [12]

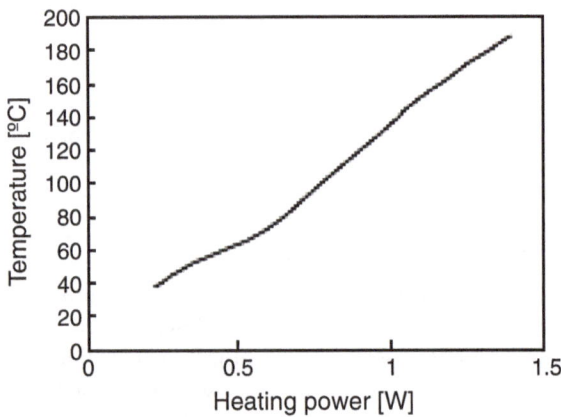

Fig. 7.8 Relation between temperature generated by microheater and heating power [12]

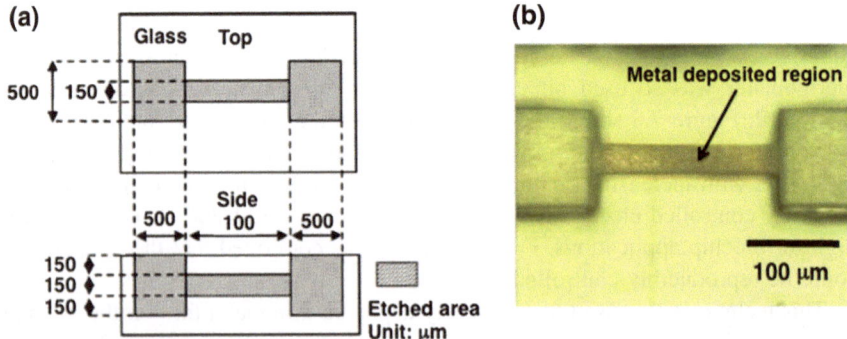

Fig. 7.9 **a** Schematic diagram of microfluidic structure fabricated in glass and **b** optical microscope image of metal-plated microfluidic channel embedded in glass [12]

thermal treatment and successive wet etching in hydrofluoric acid [2, 4]. The microfluidic structure fabricated had a 1-mm-long channel with cross-sectional dimensions of $150 \times 150 \ \mu m^2$ embedded 150 μm beneath the glass surface. The two open reservoirs had dimensions of $500 \times 500 \times 450 \ \mu m^3$. After fabricating the microfluidic channel, the femtosecond laser beam was focused on the top internal wall of the microfluidic channel in the glass and it was then scanned several times to ablate the whole area of the top wall. The glass was transparent to the femtosecond laser wavelength (775 nm) so that multiphoton absorption occurred only at the top internal wall of the microfluidic channel where the laser beam was focused. Thus, only the surface of the top internal wall was ablated. After ablation, electroless plating was performed to deposit a thin metal film on the ablated area in the microfluidic channel. Figure 7.9b shows an optical microscope image of the metal-plated microfluidic channel embedded in the glass.

A possible mechanism for selective metallization is conjectured to be as follows. When electroless plating using a mixture of a metal-ion solution and a reducing agent is performed on ablated glass, electrons generated by the reducing agent reduce the metal ions in the solution, precipitating metal atoms. As soon as the precipitated metal atoms come into contact with the glass surface, they are captured at the rough surface due to an anchoring effect, which induces strong adhesion of the deposited metal film to the substrate. In contrast, there is little anchoring at the smooth surfaces of the non-ablated regions, resulting in selective deposition of the thin metal film. Thus, surface roughening by femtosecond laser ablation is responsible for metal deposition. In fact, this technique cannot deposit metal films on fused silica due to laser ablated regions having smoother surfaces than photosensitive glass.

7.4 Fabrication of Three-Dimensional Metal Microstructures

Three-dimensional metal microstructures can be fabricated by two-photon-induced metal ion reduction using femtosecond laser direct writing [27, 31, 32]. For fabricating 3D metal microstructures, a femtosecond laser beam (wavelength: 800 nm; pulse width: 80 fs; repetition rate: 80 MHz) was tightly focused on the interface between a metal-ion solution and a glass cover slip using an oil immersion objective lens with an NA of 1.42. The focused laser beam was scanned two-dimensionally in the X–Y plane (the plane perpendicular to the laser axis) using a pair of galvanometer mirrors. After completing the X–Y plane scanning, the objective lens was translated in the Z-direction (direction of incident laser beam) using a computer-controlled motorized stage. This procedure was repeated for 3D microfabrication. Synchronizing the beam scanning with the intensity modulation induced by an electro-optical modulator enabled 3D metal microstructures to be directly fabricated using metal-ion solutions (0.2 M aqueous solution of $AgNO_3$ for silver

Fig. 7.10 Scanning electron
microscopy image of a 3D
self-standing silver gate
microstructure fabricated on a
glass substrate by two-photon
induced metal ion reduction
using femtosecond laser
direct writing [27]
(Reproduced with permission
from AIP. ©2006 by the
American Institute of
Physics)

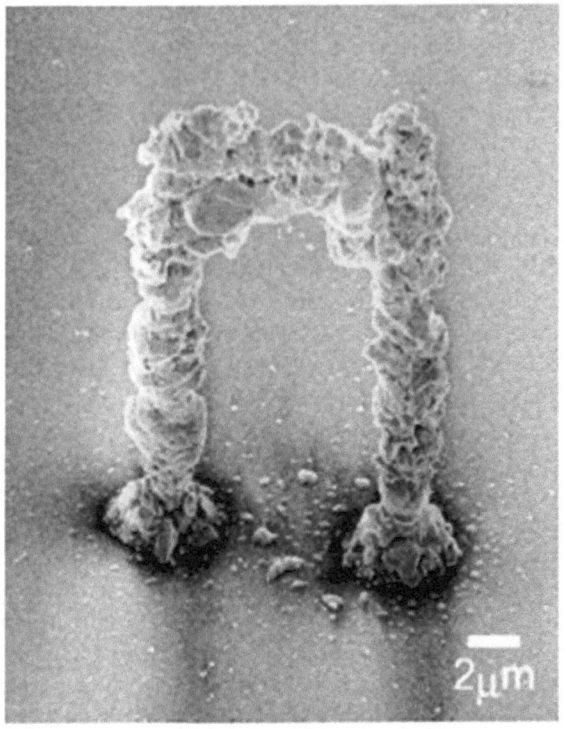

microstructures and 0.24 M aqueous solution of tetrachloroauric acid ($HAuCl_4$) for
gold microstructures). In this process, metal ions in the liquid solutions can be
directly reduced by two-photon absorption of the femtosecond laser beam, resulting
in precipitation of metal atoms that create 3D metal microstructures. The fabricated
silver microstructures had a high electrical conductivity with an average resistivity
of 5.30×10^{-8} Ωm, which is only 3.3 times greater than that of bulk silver.
Figure 7.10 shows a scanning electron microscopy image of a 3D self-standing
silver gate microstructure fabricated on a glass substrate by this technique. The
scanning speed of the laser spot during exposure was 24 μm/s. To fabricate this
structure, two poles were fabricated by scanning in the Z-direction from the bottom
(glass surface) to the top. The two pole tips were then connected by scanning the
laser beam in the X-direction. The width, height, and linewidth of the fabricated
structure were measured to be 12, 16, and 2 μm, respectively.

A similar technique was applied to create silver nanoparticles for SERS
detection of organic molecules [13] (see also Chap. 9). In this case, Rhodamine 6G
(R6G) was mixed with AgNO$_3$ solution, so that silver nanoparticles and R6G
molecules were co-deposited on the glass substrate. Raman scattering of the R6G
molecules was strongly enhanced due to localization of the electromagnetic field
by the silver nanoparticles.

Another scheme for fabricating 3D metal microstructures involves using
polyvinylpyrrolidone (PVP) films containing silver ions [31]. Figure 7.11 shows

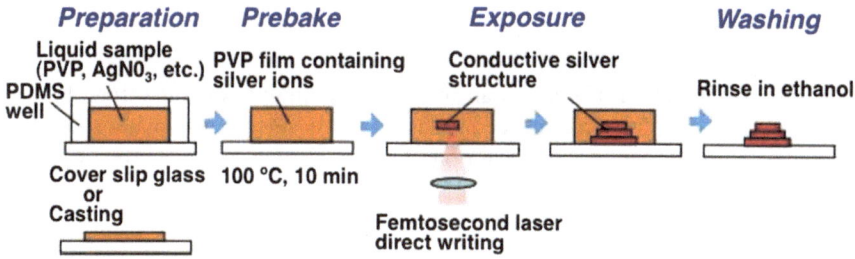

Fig. 7.11 Experimental procedure for fabricating 3D metal microstructures by femtosecond-laser-induced metal ion reduction using PVP film containing silver ions [31] (Reproduced with permission from OSA. ©2008 by the Optical Society of America)

Fig. 7.12 Laser scanning microscopy image of 3D silver microstructures produced by femtosecond-laser-induced metal ion reduction using PVP film containing silver ions [31] (Reproduced with permission from OSA. ©2008 by the Optical Society of America)

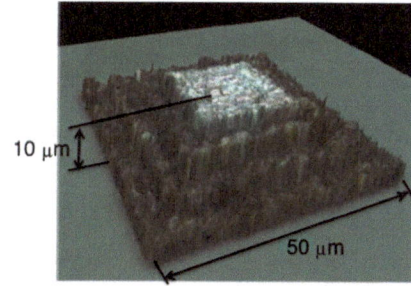

the experimental procedure for this technique. First, 1.25 g of PVP was dissolved in 50 ml of ethanol. The prepared PVP solution was mixed with $AgNO_3$ solution dissolved in deionized water (e.g., 3.8 wt % of silver nitrate in 10 ml of deionized water). To make a thin film containing silver ions, the mixed solution was spread onto a glass cover slip or stored in wells made of dimethylpolysiloxane. The cover slip coated with the mixed solution was baked in air at 110 °C for 10 min. A femtosecond laser beam (wavelength: 752 nm; pulse width: 200 fs; repetition rate: 76 MHz) was focused on the sample by an objective lens with an NA of 1.25 to fabricate metal microstructures based on the two-photon-initiated photoreduction. The 3D piezoelectric translation stage was scanned according to the 3D computer-aided design (CAD) data for fabricating 2D and 3D microstructures. After laser direct writing, the sample was soaked in ethanol for 30 min to remove the polymer matrix and it was then rinsed with deionized water. As a result, 2D and 3D metal microstructures formed on the glass substrates. The electrical resistivity of the continuous silver microstructure was measured to be 3.47×10^{-7} Ωm. Figure 7.12 shows a laser scanning microscopy image of 3D silver microstructures fabricated by this technique. The pyramidal 3D structure with a height of 10 μm was built up from the top layer because the prefabricated silver layer prevents the laser beam from being tightly focused (the laser beam was incident through the cover slip glass; i.e., from the bottom side).

7.5 Summary

Selective metallization of glass is an important technique for imparting biochips with high functionalities. Femtosecond-laser-assisted electroless plating involves irradiating a femtosecond laser beam on glass immersed in an electroless plating solution. It selectively deposits thin metal films on laser-irradiated regions by a single step, even on the internal walls of microfluidic channels. One drawback of this technique is the weak adhesion of metal films deposited on glass, although the adhesion can be improved by post annealing. Femtosecond laser modification using a $AgNO_3$ film followed by electroless plating results in higher adhesion of metal films. This technique can be used to deposit metal films on various kinds of materials. Femtosecond laser ablation followed by electroless plating also allows selective deposition of metal films with high adhesion. However, this technique is only applicable to certain materials such as photosensitive glass, since the surface roughness produced by laser ablation is responsible for metal deposition. These selective metallization techniques were demonstrated by fabricating microheaters for spatial control of the temperature in microfluidic devices and a SERS substrate for highly sensitive analysis of molecules. 3D metal microstructures with high electrical conductivities can be created by two-photon-induced metal ion reduction of a liquid or a polymer containing metal ions using femtosecond laser direct writing. This technique has not been used to fabricate biochips, but fabrication of such structures inside the microfluidic systems will enable more functionalities to be added to biochips.

References

1. Marcinkevičius A, Juodkazis S, Watanabe M et al (2001) Femtosecond laser-assisted three-dimensional microfabrication in silica. Opt Lett 26:277–279
2. Masuda M, Sugioka K, Cheng Y et al (2003) 3-D microstructuring inside photosensitive glass by femtosecond laser excitation. Appl Phys A 76:857–860
3. Bellouard Y, Said A, Dugan M et al (2004) Fabrication of high-aspect ratio, micro-fluidic channels and tunnels using femtosecond laser pulses and chemical etching. Opt Express 12:2120–2129
4. Masuda M, Sugioka K, Cheng Y et al (2004) Direct fabrication of freely movable microplate inside photosensitive glass by femtosecond laser for lab-on-chip application. Appl Phys A 78:1029–1032
5. Kiyama S, Tomita T, Matsuo S et al (2009) Laser fabrication and manipulation of an optical rotator embedded inside a transparent solid material. J Laser Micro Nanoengin 4:18–21
6. Davis KM, Miura K, Sugimoto N et al (1996) Writing waveguides in glass with a femtosecond laser. Opt Lett 21:1729–1731
7. Li ZL, Low DKY, Ho MK et al (2006) Fabrication of waveguides in Foturan by femtosecondlaser. J Laser Appl 18:320–324
8. Cheng Y, Sugioka K, Midorikawa K et al (2003) Three-dimensional micro-optical components embedded in photosensitive glass by a femtosecond laser. Opt Lett 28:1144–1146

9. Cheng Y, Tsai HL, Sugioka K et al (2006) Fabrication of 3D microoptical lenses in photosensitive glass using femtosecond laser micromachining. Appl Phys A 85:11–14

10. Wang Z, Sugioka K, Midorikawa K (2007) Three-dimensional integration of microoptical components buried inside photosensitive glass by femtosecond laser direct writing. Appl Phys A 89:951–955

11. Hanada Y, Sugioka K, Shihira-Ishikawa I et al (2011) 3D microfluidic chips with integrated functional microelements fabricated by a femtosecond laser for studying the gliding mechanism of cyanobacteria. Lab Chip 11:2109–2115

12. Hanada Y, Sugioka K, Midorikawa K (2008) Selective metallization of photostructurable glass by femtosecond laser direct writing for biochip application. Appl Phys A 90:603–607

13. Zhou Z, Xu J, He F et al (2010) Surface-enhanced Raman scattering substrate fabricated by femtosecond laser induced co-deposition of silver nanoparticles and fluorescent molecules. Jpn J Appl Phys 49:022703

14. Holland L (1963) Vacuum deposition of thin films. Chapman and Hall, London

15. Nakajima Y, Kusuyama K, Yamaguchi H et al (1992) Growth of single-crystal aluminium films on silicon substrates by DC magnetron sputtering. Jpn J Appl Phys 31:1860–1867

16. Jain A, Chi KM, Kodas TT et al (1993) Chemical vapor deposition of copper from hexafluoroacetylacetonato copper(I)—vinyltrimethylsilane deposition rates, mechanism, selectivity, morphology, and resistivity as a function of temperature and pressure. J Electrochem Soc 140:1434–1439

17. Sugioka K, Gu B, Holmes A (2007) The state of the art and future prospects for laser direct-write for industrial and commercial applications. MRS Bull 32:47–54

18. Zhang J, Sugioka K, Midorikawa K (1999) Direct fabrication of microgratings in fused quartz by laser-induced plasma-assisted ablation with a KrF excimer laser. Opt Lett 23:1486–1488

19. Hanada Y, Sugioka K, Gomi Y et al (2004) Development of practical system for laser-induced plasma-assisted ablation (LIPAA) for micromachining of glass materials. Appl Phys A 79:1001–1003

20. Adrian FJ, Bohandy J, Kim BF et al (1987) A study of the mechanism of metal-deposition by the laser-induced forward transfer. J Vac Sci Technol 5:1490–1494

21. Esrom H, Zhang J, Kogelschatz U et al (1995) New approach of a laser-induced forward transfer for deposition of patterned thin metal films. Appl Surf Sci 86:202–207

22. Glezer EN, Milosavljevic M, Huang L et al (1996) Three-dimensional optical storage inside transparent materials. Opt Lett 21:2023–2025

23. Sugioka K, Hongo T, Takai H et al (2005) Selective metallization of internal walls of hollow structures inside glass using femtosecond laser. Appl Phys Lett 86:171910

24. Xu J, Liao Y, Zeng HD et al (2007) Selective metallization on insulator surfaces with femtosecond laser pulses. Opt Express 15:12743–12748

25. Xu J, Liao Y, Zeng HD et al (2008) Mechanism study of femtosecond laser induced selective metallization (FLISM) on glass surfaces. Opt Commun 281:3505–3509

26. Baldacchini T, Pons AC, Pons J et al (2005) Multiphoton laser direct writing of twodimensional silver structures. Opt Express 13:1275–1280

27. Tanaka T, Ishikawa A, Kawata S (2006) Two-photon-induced reduction of metal ions for fabricating three dimensional electricalally conductive metallic microstructure. Appl Phys Lett 88:081107

28. Schaffer CB, García JF, Mazur E (2004) Bulk heating of transparent materials using a high-repetitionrate femtosecond laser. Appl Phys A 76:351–354

29. Watanabe W, Onda S, Tamaki T et al (2006) Space-selective laser joining of dissimilar transparent materials using femtosecond laser pulses. Appl Phys Lett 89:021106

30. Liao Y, Xu J, Cheng Y et al (2008) Electro-optic integration of embedded electrodes and waveguides in $LiNbO_3$ using a femtosecond laser. Opt Lett 33:2281–2283

31. Maruo S, Saeki T (2008) Femtosecond laser direct writing of metallic microstructures by photoreduction of silver nitrate in a polymer matrix. Opt Express 16:1174–1179

32. Cao YY, Takeyasu N, Tanaka T et al (2009) 3D metallic nanostructure fabrication by surfactant-assisted multiphoton-induced reduction. Small 5:1144–1148

Chapter 8
Integration of Microcomponents

Abstract Various microcomponents, including microelectrodes and micro-optic and microfluidic components, can be fabricated in transparent materials by femtosecond laser direct writing. This chapter describes in detail techniques for integrating different types of microcomponents on a single substrate for constructing highly functional microfluidic, photonic, and optofluidic systems and devices. Several examples are described, including integration of microlenses and waveguides for beam collimation and focusing, integration of a micro-optical ring cavity and a microfluidic chamber for creating 3D microfluidic dye lasers, integration of microelectrodes and waveguide-based Mach–Zehnder interferometer in a lithium niobate ($LiNbO_3$) crystal for constructing an optical modulator, and integration of micro-optic and microfluidic components in glass for optofluidic applications.

8.1 Introduction

As the previous chapters have shown, both microfluidic and micro-optic components can be fabricated in glass by femtosecond laser three-dimensional (3D) direct writing, making it straightforward to integrate them [1–3]. In addition, in combination with electroless plating, femtosecond laser direct writing can be used to fabricate microelectrodes in glass and crystal [4, 5]. Thus, using only a single femtosecond laser micromachining system, it is possible to integrate all fluidic, optical, and electrical functions in a single chip without assembly and packaging. Such versatility cannot be realized by any other existing technology.

The remainder of this chapter is organized into four sections: integration of micro-optical components (Sect. 8.2); fabrication of 3D microfluidic lasers in glass by integrating a micro-optical cavity and microfluidic chambers and channels (Sect. 8.3); fabrication of an optical modulator by integrating microelectrodes and a waveguide-based Mach–Zehnder interferometer (MZI) (Sect. 8.4); and integration of micro-optical and microfluidic components for optofluidic applications (Sect. 8.5).

K. Sugioka and Y. Cheng, *Femtosecond Laser 3D Micromachining for Microfluidic and Optofluidic Applications*, SpringerBriefs in Applied Sciences and Technology, DOI: 10.1007/978-1-4471-5541-6_8, © The Author(s) 2014

8.2 Integration of Free-Space Micro-optical Components and Optical Waveguides

Integration of free-space micro-optical components, such as micromirrors and microlenses, with optical waveguides can greatly enhance the performance and functionality of chemical and biological sensors. Previously, using lithography-based 2D fabrication techniques, only cylindrical lenses have been integrated with optical fibers or waveguides in polymer substrates [6, 7]. Femtosecond laser direct writing enables hybrid integration of 3D micro-optical components such as micromirrors and microspherical lenses with optical waveguides by incorporating them in a single substrate, which allows the advantages of both components to be fully exploited [8, 9].

For instance, from a practical viewpoint, a light beam guided by a waveguide integrated into a photonic or optofluidic chip often needs to be bent. To minimize the bending loss, the curvature of the waveguide should be greater than several millimeters (i.e., 5–6 mm), but this is undesirable since it increases the device size. Integration of two straight waveguides with a mirror would significantly reduce the device size [8, 10, 11]. Figure 8.1 schematically illustrates such a structure fabricated in Foturan glass. The micromirror is a hollow plate fabricated inside Foturan glass by femtosecond laser direct writing followed by thermal treatment and successive wet etching. Two waveguides were then consecutively written in glass. The first waveguide (waveguide I) was connected to the micromirror at an angle of 45° and it was connected to the second waveguide (waveguide II) at an angle of 90°. Using this structure, a sharp bend can be realized in a tight space to reduce the chip size. The bending loss at the micromirror was measured to be less than ∼0.3 dB, which includes both the reflection loss at the micro-mirror and the coupling loss from the first waveguide to the second one.

Fig. 8.1 Schematic diagram of structure fabricated in Foturan glass that integrates two optical waveguides and a micromirror [8]

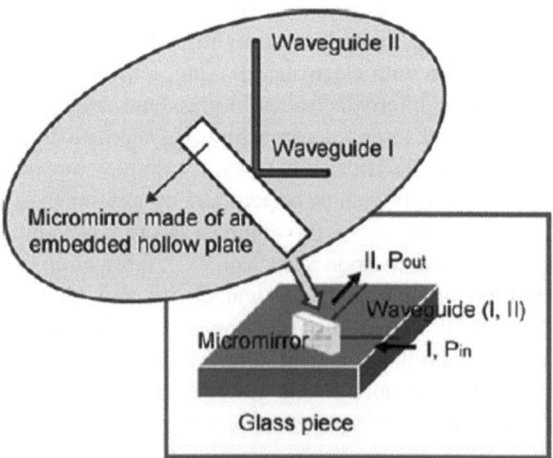

8.3 Microfluidic Dye Laser

Microfluidic dye lasers can provide coherent, wavelength-tunable radiation for optical analysis on an optofluidic chip [12–14]. Microfluidic dye lasers can easily be fabricated in glass using femtosecond laser microprocessing by straightforward integration of 3D micro-optical and 3D microfluidic components [12, 15]. Figure 8.2a shows a top view of the fabricated microfluidic laser, which has an optical microcavity consisting of four 45° micromirrors buried vertically in the glass, a horizontal microfluidic chamber embedded ~400 μm beneath the glass surface, and a microfluidic through channel that passes through the center of the microchamber. Figure 8.2b shows a micrograph of the side view of the fabricated microfluidic laser. It reveals a microchannel with an average diameter of ~80 μm and a microchamber with a thickness of ~200 μm. Figure 8.2c illustrates the optical path of the microfluidic laser. The optical cavity consists of a pair of corner

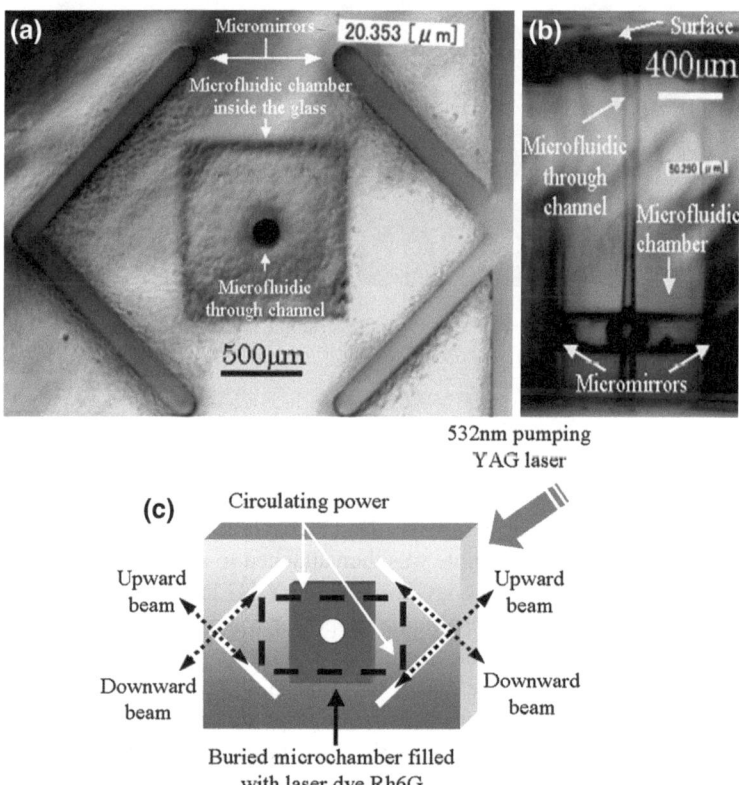

Fig. 8.2 a Optical micrograph showing *top view* of microfluidic laser; **b** optical micrograph showing *side view* of microfluidic chamber and through channel; and **c** schematic diagram showing *light path* of microfluidic laser [12]

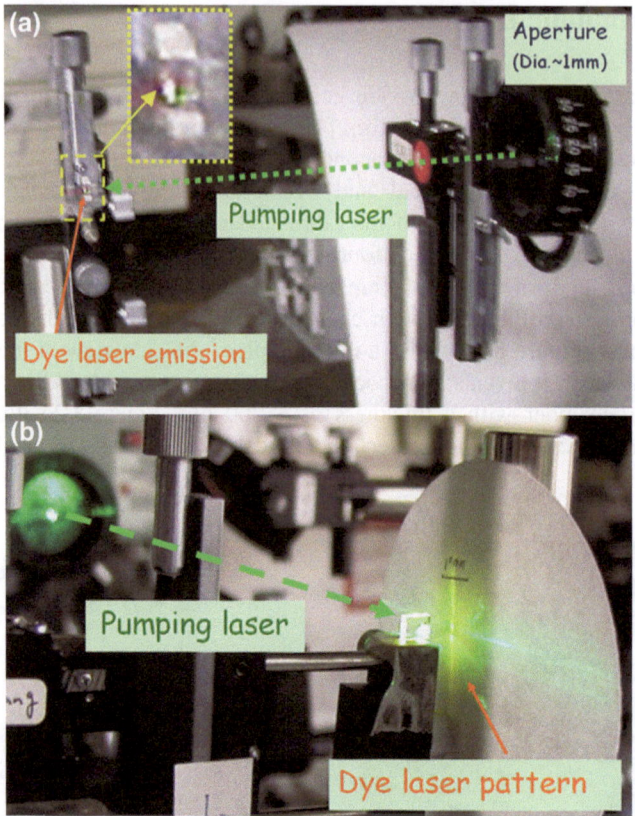

Fig. 8.3 Photographs of **a** laser emission from glass sample and **b** far-field pattern of laser beam on screen

mirrors consisting of two micromirrors on the left-hand side and two on the right-hand side. This ring cavity has a similar design to a square microcavity laser [16].

To demonstrate the functionality of the microfluidic dye laser, the microfluidic chamber was filled with the laser dye Rh6G dissolved in ethanol ($\sim$0.02 mol/l) using a syringe needle. The sample was then attached to an optical alignment stage and pumped by a pulsed, frequency-doubled Nd:YAG laser (pulse duration: 5 ns; repetition rate: 15 Hz). When the pumping power was increased above the lasing threshold, light emission was clearly observed from the output of the micro-dye laser, as shown in Fig. 8.3a. A screen was placed approximately 8 mm from the output end of the structure to enable a photograph to be obtained of the far-field emission pattern, as shown in Fig. 8.3b. Two laser beams emitted tangentially from the internal surfaces of the two 45° mirrors were simultaneously observed on the screen; one propagated upward and the other propagated downward. The laser beams were well confined in the direction perpendicular to the plane of the optical cavity; however, the emissions lacked directionality in the plane of the optical

Fig. 8.4 Emission spectra of microfluidic laser at pumping fluences of **a** 0.46, **b** 1.66, and **c** 4.49 mJ/cm². The peaks centered at 532 nm originate from scattered pumping light

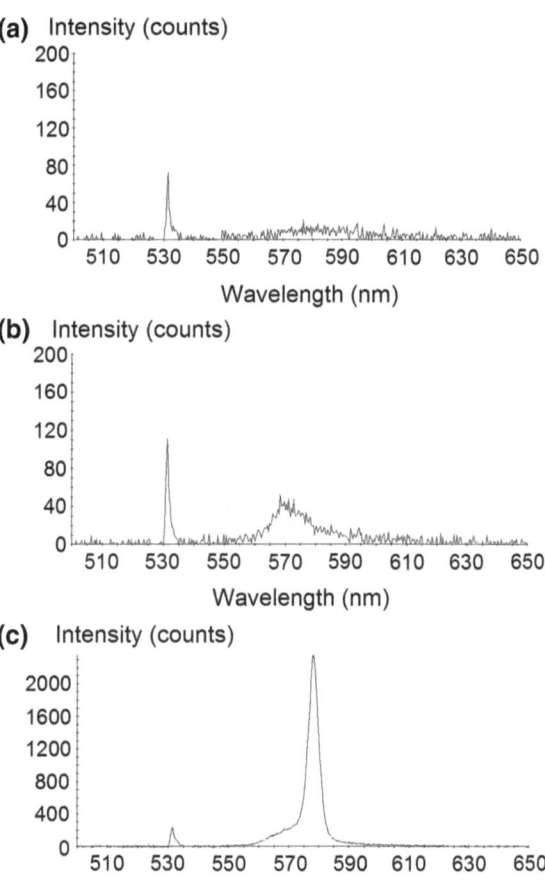

cavity. This is a common problem for microcavity lasers since light leaks from the optical microcavity due to scattering from the sidewalls, which is essentially a random process.

Figure 8.4 shows emission spectra of the microfluidic laser measured at different pumping energies. The detector head of the spectrometer (USB2000, Ocean Optics, Inc.) was placed near the output of the microfluidic laser to collect the light from the downward-propagating beam. After each measurement, a power meter (Lasermate, Coherent, Inc.) was used to measure the corresponding average power of the pumping laser. As shown in Fig. 8.4, when a low pumping fluence of 0.46 mJ/cm² was applied, spontaneous emission with a wide spectrum was observed. Lasing action commenced near a pumping fluence of 1.66 mJ/cm²; at this fluence the spectrum started to become narrower, as shown in Fig. 8.4b. Further increasing the pumping fluence rapidly increased the output power of the microfluidic laser and reduced its bandwidth. A typical emission spectrum with a center wavelength of ∼578 nm of the microfluidic laser under high pumping energy of 4.49 mJ/cm² was measured, as shown in Fig. 8.4c. In addition, the average output power was measured to be ∼10 μW at a repetition rate of 15 Hz, corresponding to a pulse energy of approximately 1 μJ from the microfluidic laser.

Furthermore, 3D femtosecond laser microprocessing can also be used to fabricate an array of microfluidic dye lasers arranged in multilayered configurations [15]. By arranging two microfluidic chambers serially in glass (see Fig. 8.5a), a microfluidic dual-color laser was fabricated to simultaneously produce an array of two laser emissions with different wavelengths using only a single pumping laser. After filling the two chambers with the two laser dyes Rhodamine 640 and Rh6G and pumping using a frequency-doubled Nd:YAG laser, two laser beams with different wavelengths were simultaneously emitted from different positions inside the glass. Figure 8.5b shows a measured spectrum of this dual-color laser. It reveals the two lasing wavelengths of Rh6G and Rhodamine 640 centered at 568 and 618 nm, respectively. Since microchamber arrays with more chambers can be fabricated and each chamber can be filled with a different laser dye, it is potentially possible to fabricate a tiny laser with a broad wavelength range for chemical or biological analysis.

Fig. 8.5 a 3D schematic diagram of *dual-color* microfluidic laser. **b** Measured spectrum of *dual-color* laser using Rhodamine 640 and Rh6G dyes as lasing media [15]

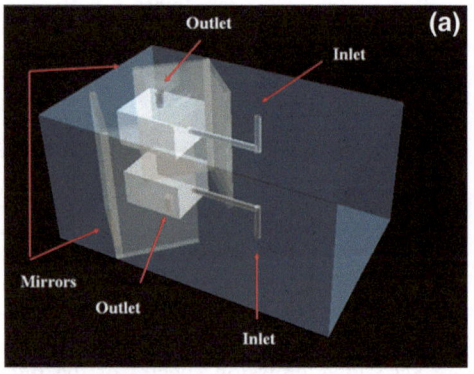

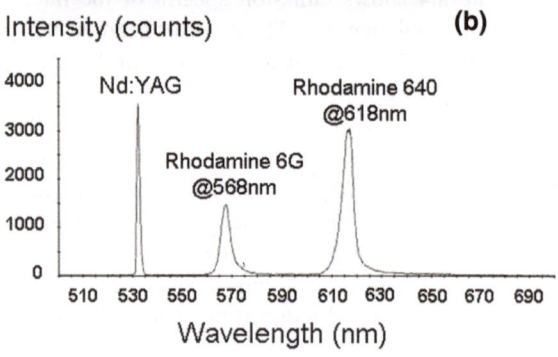

8.4 Mach–Zehnder Optical Modulator

To develop monolithic and compact multifunctional systems, it is often desirable to integrate functions other than those provided by optical and fluidic components in a single chip. Of them, an electrical conductivity is particularly attractive since it enables active systems/devices to be fabricated. By selectively metallizing surfaces of dielectric materials by femtosecond laser processing, microelectric circuits can be formed on the surfaces of various transparent materials such as glass, crystals, and polymers [4, 5]. The basic procedures include selective modification of insulator surfaces by femtosecond laser direct writing and successive electroless plating in the modified regions, as described in detail in Chap. 7.

Microelectric circuits can be monolithically integrated into microfluidic or photonic chips. For example, LiNbO$_3$ is a ferroelectric material with a large optical nonlinearity. Electro-optical (EO) integration on LiNbO$_3$ substrates has attracted significant attention, leading to active devices such as EO switches, optical modulators, and EO-tuned quasi-phase-matched devices [17–19]. Using a femtosecond laser, such EO integration can be realized by a single exposure process followed by batch electroless plating [20].

Figure 8.6 schematically depicts a Mach–Zehnder interferometer (MZI) EO modulator. The MZI consisted of a waveguide-based beam splitter and a coupler written in a LiNbO$_3$ substrate. To produce thermally stable waveguides, we wrote two parallel lines with a small separation in the low repetition rate regime; this produces a guiding region between the double lines (typically termed a type II waveguide) [21]. Compared with waveguides that guide light in an irradiated

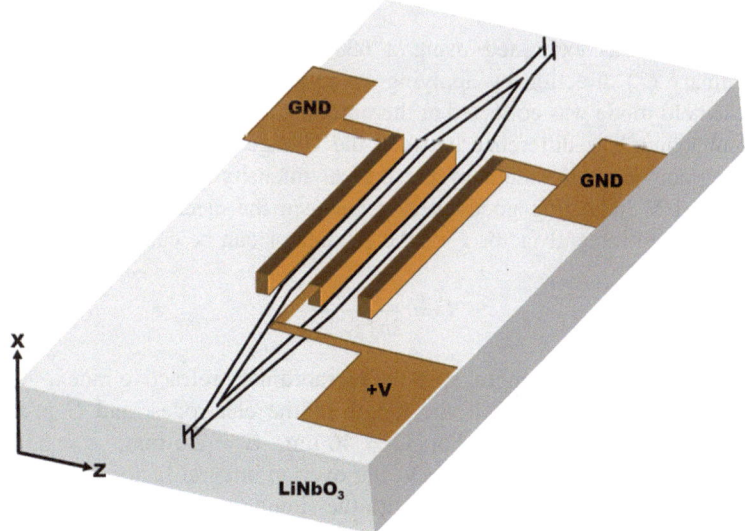

Fig. 8.6 Schematic layout of Mach–Zehnder EO modulator [20]

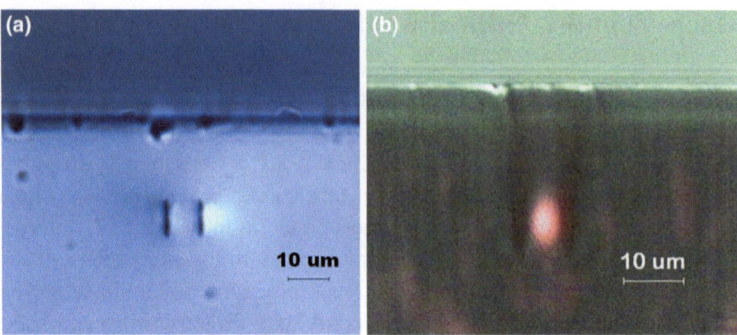

Fig. 8.7 **a** Microscope image of end facet of MZI and **b** near-field intensity distribution measured at exit of MZI [20]

region (typically termed as type I waveguides), type II waveguides preserve the nonlinearity of the bulk crystal in the guiding region [22]. Figure 8.7a shows a microscope image of the end facet of the MZI. A 633 nm He–Ne laser beam was coupled into the entrance of the MZI by a 20x microscope objective. The near-field beam profile at the exit is shown in Fig. 8.7b, revealing a single-mode distribution with a horizontal size of ~ 4 μm and a vertical size of ~ 6 μm ($1/e^2$ intensity).

The electrodes realized by femtosecond laser direct writing are deeply buried in the LiNbO$_3$ substrate, as shown in Fig. 8.8a. A numerical simulation based on the finite element method was used to analyze the electric field between the embedded electrodes. Figure 8.8b shows the calculated potential distribution. It reveals that the electric field across the waveguide is nearly uniform and is almost confined in the horizontal (Z) direction, leading to greater EO overlap than conventional planar electrodes.

The device was examined using a 633 nm He–Ne laser polarized in the extraordinary (Z) direction by applying a varying DC voltage to the electrodes. The near-field mode was collected at the output end of MZI. Figure 8.9 shows the results obtained. The difference between the voltages corresponding to the adjacent maximum and minimum of the optical intensity output was measured to be $\sim 19 \pm 1$ V indicating good overlap between the electric and optical fields. The EO overlap integral factor Γ of the modulator can be calculated using

$$\Gamma = \frac{\lambda G}{2n_e^3 \gamma_{33} V_\pi L} \qquad (8.1)$$

where λ is the laser wavelength, n_e is the extraordinary refractive index, γ_{33} is the EO coefficient of LiNbO$_3$, L is the length of the electrodes, and G is the gap between the electrodes [23]. For G = 46 μm, $G = 2.6$ mm, $\lambda = 632.8$ nm, $n_e = 2.2$, and $\gamma_{33} = 2.9 \times 10^{-11}$ m/V, the overlap integral factor is calculated to be $\Gamma = 0.95$. This value agrees well with the simulation result and is higher than that achievable for electrodes fabricated on the substrate surface by conventional photolithography.

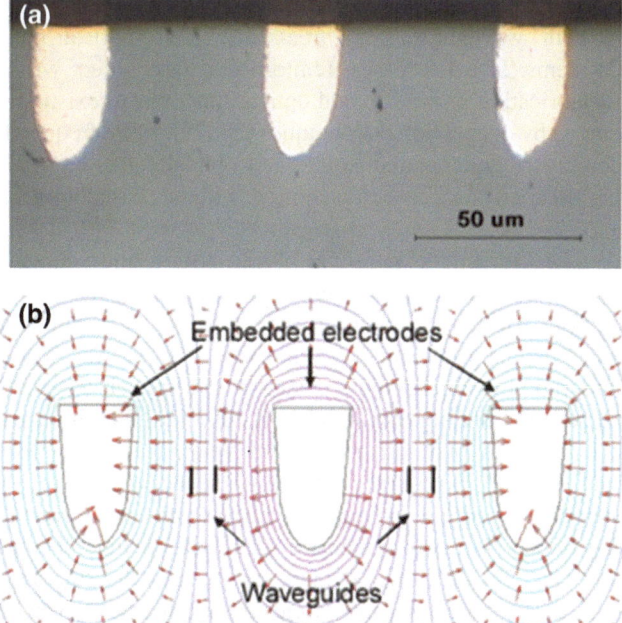

Fig. 8.8 **a** End view of optical micrographs of embedded electrodes and optical waveguides. **b** Plot of equipotential contours of embedded electrodes [20]

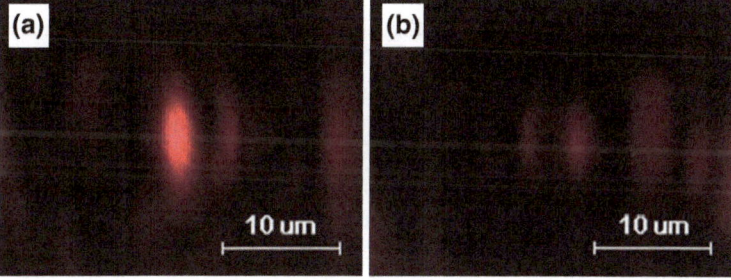

Fig. 8.9 Near-field intensity distributions at exit of EO modulator for dc voltages of **a** 0 and **b** 19 V [20]

8.5 Optofluidic Systems

The unique ability of femtosecond laser direct writing to simultaneously create fluidic and optical functions in glass opens up new avenues for fabricating various optofluidic microchips for biological analysis. Currently, there are three main approaches for fabricating optofluidic systems and devices by femtosecond laser processing: incorporating optical waveguides in microfluidic systems;

incorporating free-space micro-optical components in microfluidic systems; and incorporating both waveguides and free-space micro-optical components in microfluidic systems. In a few cases, femtosecond laser direct writing has been used as a post-fabrication process to add optical functions to existing microfluidic systems fabricated by conventional techniques [24, 25]. More frequently, complete optofluidic devices with integrated fluidic and optical functions are obtained by just using femtosecond laser micromachining without assembling or packaging processes.

Waveguides are important building blocks for various functional optical components such as beam couplers and splitters, Bragg gratings, optical ring cavities, and MZIs. Thus, incorporating waveguides in microfluidic systems can greatly enhance device functionality. An early attempt was made to realize monolithic integration of microfluidic channels and optical waveguides in fused silica by femtosecond laser micromachining in 2003 [26], but this area of research did not really begin to take off until 2006 when techniques for fabricating microfluidic systems and waveguides in fused silica became mature. Subsequently, optofluidic devices based on waveguide–microchannel integration have been intensively developed. Systems demonstrated to date include chemical and biological sensors [24, 27, 28], flow cytometers [3, 25, 29, 30], and optofluidic tweezers [31, 32]. The applications of fabricated optofluidic devices are reviewed in Chap. 9.

The example shown in Fig. 8.10 clearly demonstrates the unique 3D capability of femtosecond laser micromachining for out-of-plane integration of a waveguide-based MZI and a microfluidic channel [24]. In this specific case, the two arms of the interferometer lie in a plane tilted at 7° to the substrate plane. Thus, the reference arm is constructed above the microchannel and the sensing arm and the microchannel intersect at right angles. This orthogonal intersection of the waveguide and the microchannel offers a high spatial resolution for refractive index sensing comparable to the diameter of the waveguide mode (~ 11 μm). Conventional MZI refractive index sensors fabricated by planar fabrication techniques frequently employ the evanescent field of the waveguide mode to probe analytes [33]. In such cases, highly localized sensing is difficult to achieve because long interaction lengths are required to ensure sufficiently high sensitivities.

Figure 8.10b, c shows microscope images of the microfluidic channel and the two arms of the MZI. The microchannel is embedded 500 μm beneath the surface and has a height of ~ 50 μm and a width of ~ 110 μm. The waveguides have nearly circular cross sections with a diameter of ~ 15 μm. The reference arm was written 20 μm above the microchannel (Fig. 8.10b), whereas the sensing arm was written in the horizontal plane intersecting the center of the microchannel (Fig. 8.10c). Since the MZI has two unbalanced arms, a spectral interference fringe can be obtained at the output end using a wavelength-tunable laser source. When the refractive index changes in the microfluidic channel, the propagation speed of a light beam in the sensing arm is altered, which causes a phase shift in the recorded spectral interference fringe. Quantitative determination of the

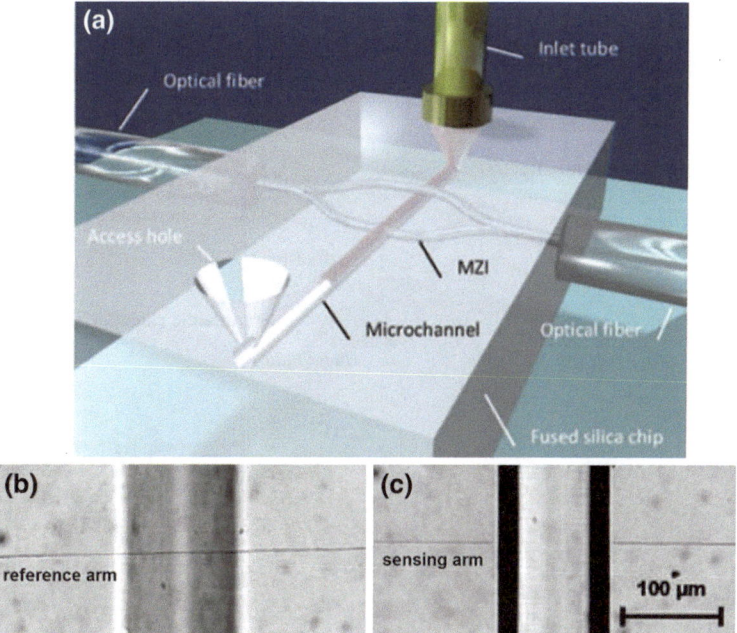

Fig. 8.10 a Schematic of a refractive index sensor fabricated by femtosecond laser direct writing. The two arms of the MZI lie in a tilted plane such that the reference arm lies above the microchannel and only the sensing arm intersects the microchannel. **b** Optical micrograph of reference arm passing over the microchannel. **c** Optical micrograph of sensing arm intersecting the microchannel [24] (Reproduced with permission from RSC. ©2010 by the Royal Society of Chemistry)

refractive index variation was achieved by measuring the phase shift of the spectral interference fringe [24], as is described in detail in Chap. 9.

Optofluidic applications can further benefit from integration of both waveguides and micro-optical components in microfluidic structures in glass materials by fully exploiting the capability of femtosecond laser direct writing. In this case, special care needs to be taken because waveguides must be written in systems after fabricating the micro-optical and microfluidic components (which can be integrated simultaneously). This is because to produce high-quality micro-optical components in either Foturan or fused silica, post-annealing processes are required to smooth the etched surfaces. If optical waveguides are written in the substrates prior to annealing, they may be destroyed during the high-temperature annealing. Although this two-step process somewhat increases the complexity of the fabrication process, it significantly enhances device functionality and performance, as evidenced by the following example.

Figure 8.11a shows an integrated optofluidic sensor that is capable of performing both absorption and fluorescence spectroscopic detection of liquid samples [9]. In this device, a 6-mm-long optical waveguide is connected to a microfluidic reservoir

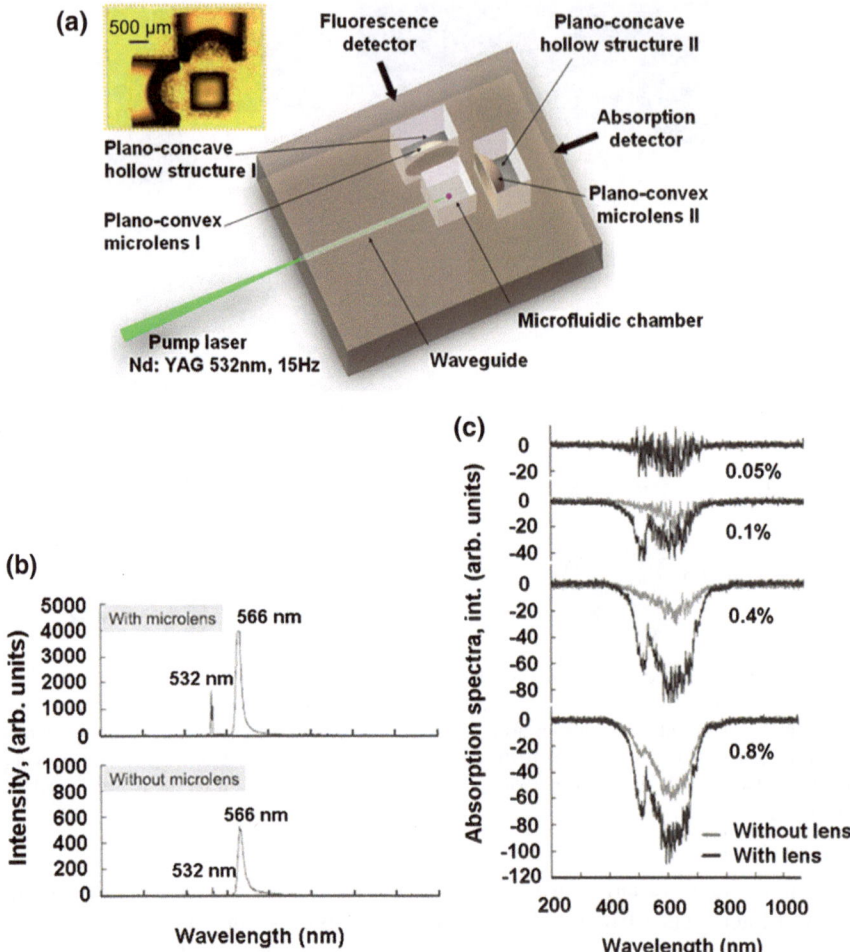

Fig. 8.11 **a** Schematic of optofluidic microchip in which micro-optical components, such as micro planoconvex lenses and an optical waveguide, are integrated with a microfluidic chamber in a single glass chip. Inset (*upper left corner*): *top view* of microchip fabricated by femtosecond laser direct writing. **b** Emission spectra from laser dye Rh6G pumped by a frequency-doubled Nd:YAG laser obtained using the microchip with (*top*) and without (*bottom*) the microlens. **c** Optical absorption spectra of *black* ink diluted in water at different concentrations in the microchamber of a microchip integrated with (*black lines*) and without (*gray lines*) a microlens. The absorption spectra were obtained by subtracting the reference spectrum (i.e., the transmission spectrum without *black* ink) from the measured transmission spectra [9]

that has dimensions of $1.0 \times 1.0 \times 1.0$ mm^3. The waveguide is used to transfer either the fluorescence excitation light from a frequency-doubled YAG laser or a broadband beam from a white lamp for absorbance measurements to a liquid sample confined in the microreservoir. Since both the fluorescence light emitted from the excited sample and the beam exiting the waveguide will propagate in space with highly divergent angles, the signal collection efficiency will be quite low. To compensate for this loss, two microspherical lenses (also with in-plane geometries and a radius of curvature of 0.75 mm) were fabricated to the side of and behind the microreservoir to improve the collection efficiencies of the fluorescence emission and the transmission light, respectively. To experimentally characterize the performance of this device, solutions of Rh6G dye and of water with different concentrations of black ink were respectively used as fluorescent and absorbing samples. As shown in Fig. 8.11b, c, the two microlenses enhanced the detection efficiencies of fluorescence and absorption spectroscopic measurements by factors of 8 and 3, respectively.

8.6 Summary

Femtosecond lasers have been used to fabricate microelectrodes and microfluidic and micro-optical components in glass in a compatible manner. Integration of these basic elements in single substrates by femtosecond laser direct writing offers the possibility of creating many innovative and extraordinary microsystems and devices. Early examples along this direction include integration of a micro-optical mirror and two waveguides for creating a sharp bend in a tight space for reducing the chip size, integration of an optical ring cavity and microfluidic chambers for constructing microfluidic lasers, integration of a waveguide-based MZI and microelectrodes for constructing an optical modulator, and integration of waveguides and/or free-space optics with microfluidic components for constructing optofluidic sensors based on refractive-index sensing as well as absorption and fluorescence spectroscopy. Optofluidic microsystems and devices fabricated by femtosecond lasers are increasingly being used in biological and biomedical research, as is discussed in Chap. 9.

References

1. Sugioka K, Cheng Y (2012) Femtosecond laser processing for optofluidic fabrication. Lab Chip 12:3576–3589
2. Osellame R, Hoekstra HJWM, Cerullo1 G et al (2011) Femtosecond laser microstructuring: an enabling tool for optofluidic lab-on-chips. Laser Photonics Rev 5:442–463
3. Schaap A, Rohrlack T, Bellouard Y (2012) Optical classification of algae species with a glass lab-on-a-chip. Lab Chip 12:1527–1532
4. Xu J, Liao Y, Zeng HD et al (2007) Selective metallization on insulator surfaces with femtosecond laser pulses. Opt Express 15:12743–12748
5. Hanada Y, Sugioka K, Midorikawa K (2008) Selective metallization of photostructurable glass by femtosecond laser direct writing for biochip application. Appl Phys A 90:603–607
6. Camou S, Fujita H, Fujii T (2003) PDMS 2D optical lens integrated with microfluidic channels: principle and characterization. Lab Chip 3:40–45

7. Wang Z, El-Ali J, Engelund M (2004) Measurements of scattered light on a microchip flow cytometer with integrated polymer based optical elements. Lab Chip 4:372–377
8. Wang Z, Sugioka K, Hanada Y et al (2007) Optical waveguide fabrication and integration with a micro-mirror inside photosensitive glass by femtosecond laser direct writing. Appl Phys A 88:699–704
9. Wang Z, Sugioka K, Hanada Y et al (2007) Three-dimensional integration of microoptical components buried inside photosensitive glass by femtosecond laser direct writing. Appl Phys A 89:951–955
10. Cheng Y, Sugioka K, Midorikawa K (2006) Freestanding optical fibers fabricated in a glass chip by femtosecond laser micromachining for lab-on-a-chip application: erratum. Opt Express 14:11910
11. Li LX, Nordin G, English J et al (2003) Small-area bends and beamsplitters for lowindex-contrast waveguides. Opt Express 11:282–290
12. Cheng Y, Sugioka K, Midorikawa K et al (2004) Microfluidic laser embedded in glass by three-dimensional femtosecond laser microprocessing. Opt Lett 29:2007–2009
13. Helbo B, Kristensen A, Menon A (2003) A micro-cavity fluidic dye laser. J Micromech Microeng 13:307–311
14. Li ZY, Zhang ZY, Emery T et al (2006) Single mode optofluidic distributed feedback dye laser. Opt Express 14:696–701
15. Cheng Y, Sugioka K, Midorikawa K (2005) Microfabrication of 3D hollow structures embedded in glass by femtosecond laser for Lab-on-a-chip applications. Appl Surf Sci 248:172–176
16. Moon HJ, Chough YT, An K (2000) Cylindrical microcavity laser based on the evanescent-wave-coupled gain. Phys Rev Lett 85:3161–3164
17. Ramer OG (1982) Integrated optic electrooptic modulator electrode analysis. IEEE J Quant Electron 18:386–392
18. Wooten EL, Kissa KM, Yi-Yan A et al (2000) A review of lithium niobate modulators for fiber-optic communications systems. IEEE J Sel Top Quant 6:69–82
19. Lu YQ, Wan ZL, Wang Q et al (2000) Electro-optic effect of periodically poled optical superlattice $LiNbO_3$ and its applications. Appl Phys Lett 77:3719–3721
20. Liao Y, Xu J, Cheng Y et al (2008) Electro-optic integration of embedded electrodes and waveguides in $LiNbO_3$ using a femtosecond laser. Opt Lett 33:2281–2283
21. Burghoff J, Grebing C, Nolte S et al (2006) Efficient frequency doubling in femtosecond laser-written waveguides in lithium niobate. Appl Phys Lett 89(3):081108
22. Thomas J, Heinrich M, Burghoff J et al (2007) Femtosecond laser-written quasi-phase-matched waveguides in lithium niobate. Appl Phys Lett 91(3):151108
23. Binh LN (2006) Lithium niobate optical modulators Devices and applications. J Cryst Growth 288:180–187
24. Crespi A, Gu Y, Ngamsom B et al (2010) Three-dimensional Mach-Zehnder interferometer in a microfluidic chip for spatially-resolved label-free detection. Lab Chip 10:1167–1173
25. Applegate RW, Squier J, Vested T et al (2006) Microfluidic sorting system based on optical waveguide integration and diode laser bar trapping. Lab Chip 6:422–426
26. Bellouard Y, Said AA, Dugan M et al (2003) Monolithic three-dimensional integration of micro-fluidic channels and optical waveguides in fused silica. Proc Mater Res Soc Fall Meet Symp A (Mater Res Soc) 782:63–68
27. Maselli V, Grenier JR, Ho S et al (2009) Femtosecond laser written optofluidic sensor: Bragg grating waveguide evanescent probing of microfluidic channel. Opt Express 17:11719–11729
28. Vazquez RM, Osellame R, Nolli D et al (2009) Integration of femtosecond laser written optical waveguides in a lab-on-chip. Lab Chip 9:91–96
29. Kim M, Hwang DJ, Jeon H et al (2009) Single cell detection using a glass-based optofluidic device fabricated by femtosecond laser pulses. Lab Chip 9:311–318
30. Schaap A, Bellouard Y, Rohrlack T et al (2011) Optofluidic lab-on-a-chip for rapid algae population screening. Biomed Opt Express 2:658–664

31. Bragheri F, Ferrara L, Bellini N et al (2010) Optofluidic chip for single cell trapping and stretching fabricated by a femtosecond laser. J Biophotonics 3:234–243
32. Bellini N, Vishnubhatla KC, Bragheri F et al (2010) Femtosecond laser fabricated monolithic chip for optical trapping and stretching of single cells. Opt Express 18:4679–4688
33. Heideman RG, Lambeck PV (1999) Remote opto-chemical sensing with extreme sensitivity: design, fabrication and performance of a pigtailed integrated optical phase-modulated Mach-Zehnder interferometer system. Sens Actuat B 61:100–127

Chapter 9
Applications of Biochips Fabricated by Femtosecond Lasers

Abstract The ability of femtosecond laser processing to simultaneously fabricate three-dimensional microfluidic, micro-optical, microelectronic, and micromechanic components inside glass microchips provides great advantages over conventional fabrication techniques for fabricating various biochips. This chapter introduces applications of biochips fabricated by femtosecond laser processing to biosensing based on surface-enhanced Raman scattering spectroscopy, efficient mixing of fluids, single cell detection, manipulation and sorting of cells, concentration analysis of liquid samples, and detection and elucidation of the functions of microorganisms and bacteria.

9.1 Introduction

As described in previous chapters, femtosecond lasers allow us to fabricate various functional microcomponents in glass including three-dimensional (3D) microfluidic [1–4], micro-optical [5–7], microelectronic [8–10], and micromechanic components [11, 12]. Further, Chap. 8 has demonstrated that a single femtosecond laser micromachining system can integrate these microcomponents in a single glass chip without assembly and packaging [13–18]. Such integration provides many advantages over existing technologies such as soft lithography and semiconductor processing based on photolithography for fabricating biochips such as microfluidic and optofluidic systems, lab-on-a-chip devices, and micro-total analysis systems (μ-TAS).

This chapter introduces applications of biochips fabricated by femtosecond laser processing, which involve biosensing based on surface-enhanced Raman scattering (SERS) spectroscopy [19], efficient mixing of fluids [20], single cell detection [21], manipulation and sorting of cells [22–26], concentration analysis of liquid samples [18, 27–29], and detection and elucidation of the functions of microorganisms and bacteria [28, 30–33].

K. Sugioka and Y. Cheng, *Femtosecond Laser 3D Micromachining for Microfluidic and Optofluidic Applications*, SpringerBriefs in Applied Sciences and Technology, DOI: 10.1007/978-1-4471-5541-6_9, © The Author(s) 2014

9.2 Biosensing Based on Surface-Enhanced Raman Scattering Spectroscopy

Metallization of glass substrates by femtosecond laser photoreduction using silver nitrate ($AgNO_3$) solution was applied to highly sensitive sensing of molecules based on SERS [19]. The femtosecond laser beam was focused on the glass cover slip surface that was in contact with the $AgNO_3$ solution. Silver ions in the $AgNO_3$ solution were then reduced so that silver thin films were deposited on the glass surface, as shown in Fig. 9.1a. It shows that silver particles with diameters ranging from tens of nanometers to a few micrometers were partially created. To demonstrate sensing based on SERS, 10 ml of Rhodamine 6G (R6G) solution diluted with methanol (10^{-7} M) was dropped on the silver film as a sample. The solvent was evaporated and a watermark area of less than 1/4 of the original area remained on the cover slip. After 3 min, the cover slip was rinsed with distilled water and allowed to dry naturally. Figure 9.1b shows a SERS spectrum of R6G obtained using an integration time of 1 s after the background has been subtracted (the inset shows the original spectrum before subtracting the background). Weak but distinct Raman bands, especially for the aromatic C-C stretching vibrations in the spectral range 1300–1650 cm^{-1}, are observed with an intense fluorescence background. The Raman signal originates from R6G molecules attaching to the SERS-active spots, while the fluorescence comes from dissociative molecules. This result indicates that the deposited silver nanostructures give rise to a SERS effect that permits sensitive detection. The detection sensitivity (10^{-7} m) for SERS realized using this microchip is comparable to previous experimental results [34, 35]. However, the Raman enhancement factor (EF) is relatively low because only a few R6G molecules were effectively adsorbed on the SERS active spots by diffusion in the present scheme. Unlike silver colloidal substrates, the lack of electrostatic

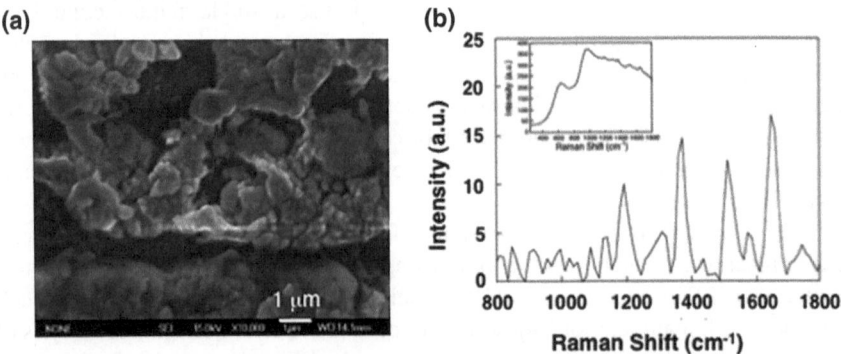

Fig. 9.1 a SEM image of silver film deposited on glass chip by irradiation in $AgNO_3$ solution. **b** SERS spectrum of R6G molecules (10^{-7} m) adsorbed on silver nanoparticles after background subtraction (integration time: 1 s). The *inset* shows the original SERS spectrum [19]

interactions between the molecules and the silver film prevents them from getting closer and hence limits the total EF.

To enhance the EF, R6G molecules were embedded by mixing R6G with AgNO$_3$ solution and the mixed solution was irradiated by a femtosecond laser beam. Organic R6G molecules were then co-deposited on the slip surface together with the silver nanoparticles. Consequently, the EF of a sensor with embedded molecules had an EF that was 40 higher than that without embedded molecules. It could detect R6G molecules with a concentration as low as 10^{-8} M.

9.3 Fluid Mixing

Microfluidic channels with diameters in the range of hundreds of nanometers to hundreds of micrometers usually generate laminar fluid flow since such dimensions have low Reynolds numbers, as shown in Fig. 9.2a [36]. Therefore, efficient mixing of two or more fluids in microfluidic systems is a critical issue in biochip applications. Chaos theory predicts that a passive micromixer based on Baker's transformation will provide ideal mixing [37]. A passive microfluidic mixer consisting of symmetrical 3D units has been theoretically proposed (see Fig. 9.2b) [38]. However, since it has a complex 3D geometry, it cannot be fabricated by conventional planar fabrication processes. Ablation induced by femtosecond laser direct writing of mesoporous glass immersed in water followed by post-annealing can be used to fabricate microfluidic structures with nearly unlimited lengths and arbitrary geometries [4, 20, 39] (see Chap. 4). Using this technique, a passive microfluidic mixer has been constructed in mesoporous glass, as shown in

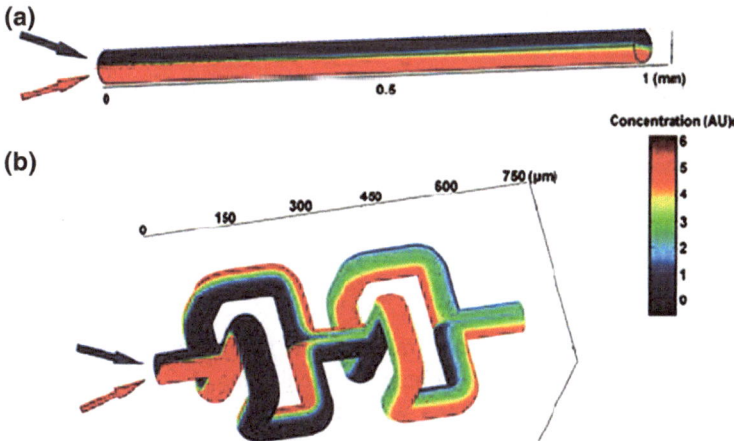

Fig. 9.2 Numerical simulations of fluid mixing in (**a**) simple 1D microfluidic channel and (**b**) 3D microfluidic mixer theoretically proposed based on Baker's transformation [20]

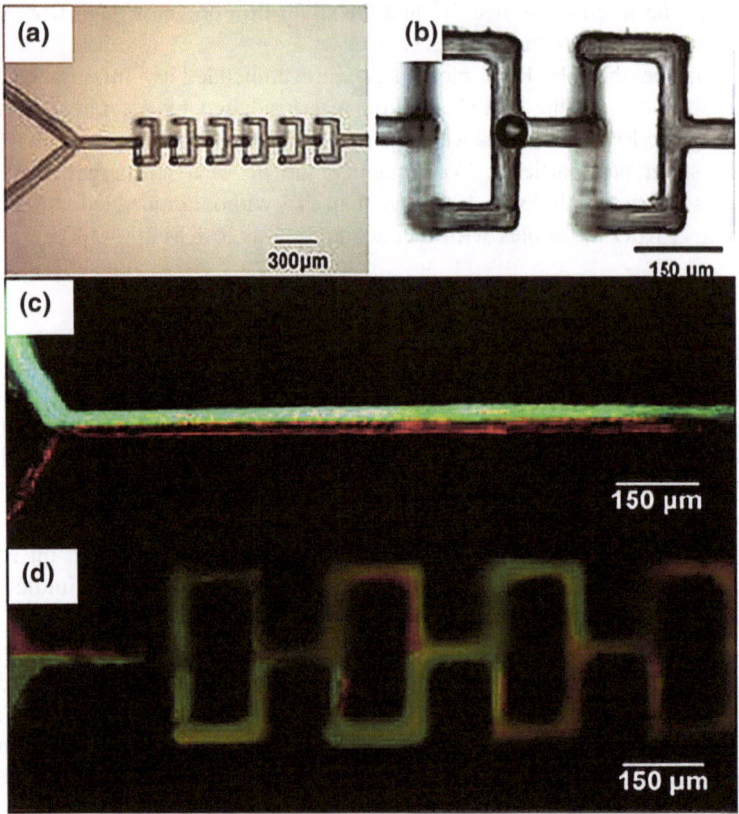

Fig. 9.3 **a** *Top-view* micrograph of a 3D microfluidic mixer fabricated in glass. **b** *Close-up view* of mixing units. **c** Mixing experiment performed in a simple 1D microfluidic channel. **d** Mixing performance of fabricated 3D microfluidic mixer [20]

Fig. 9.3a, b [20]. The fabricated microfluidic structure exhibits a higher mixing efficiency of two fluids (fluorescein sodium and Rhodamine B) than its 1D counterpart (see Fig. 9.3c, d).

9.4 Single Cell Detection

Detection of single cells is very important in many fields including biology, chemistry, food safety, and environmental monitoring. Optofluidic systems fabricated by femtosecond laser direct writing consisting of a microfluidic channel integrated with optical waveguides that intersect the microfluidic channel at right angles (see Fig. 9.4a), allow rapid detection of single cells [21]. The microfluidic channel in this system was first fabricated in fused silica by femtosecond laser

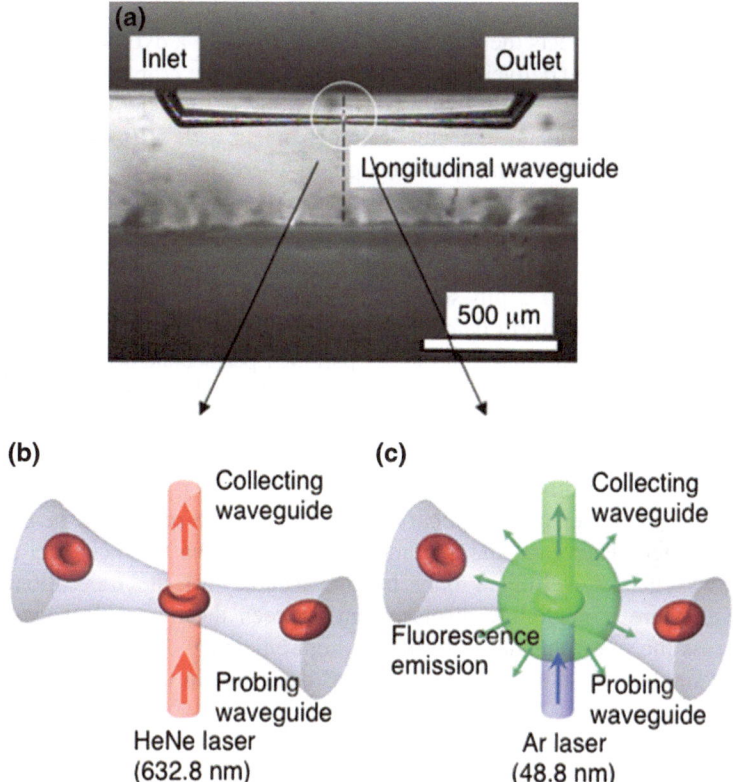

Fig. 9.4 **a** Optical transmission image of an integrated optofluidic device fabricated for detecting single cells. *The dashed line* indicates the location of the integrated longitudinal waveguide. Schematic illustration of cell detection experiment using **b** transmission intensity change and **c** fluorescence emission [21] (Reproduced with permission from RSC. ©2009 by the Royal Society of Chemistry)

direct writing followed by wet chemical etching in dilute hydrofluoric acid [1]. Longitudinal optical waveguides were then written by refractive index modification induced by femtosecond laser direct writing [5]. They were normal to the top surface of the glass substrate so that they perpendicularly intersect the prefabricated microchannels. The microchannel diameter at the point where they intersect the optical waveguides could be varied between 5 and 100 μm by adjusting the etch time and the laser scanning parameters. The dashed line in Fig. 9.4a indicates the position of the written optical waveguides. The fabricated optofluidic system was used to detect single red blood cells (RBCs) in diluted human blood. Figure 9.4b schematically illustrates the passive optical detection setup used to measure the transmitted light intensity and Fig. 9.4c schematically depicts fluorescence emission detection. The former approach involves sensing the intensity change of the probing waveguide-delivered He–Ne laser light (632.8 nm) induced by the refractive index change caused by a cell flowing in the channel. In contrast, the latter approach detects the fluorescence emission from a dyed RBC excited by Ar laser light (488 nm) delivered by the optical waveguide probe (bottom side). In both cases, the attenuated or fluorescence light is delivered by the collecting waveguide (top side) to the detector. Microchannels with 5 μm-diameter necks (i.e., the microchannel diameter at the point where they intersect the optical waveguides), whose diameters are slightly smaller than a RBC (6–8 μm) generated a sharp, constant, and unambiguous signal for cell detection in both detection schemes. This is because healthy RBCs can squeeze into narrow microchannels with diameters down to 2 μm so that the small neck diameter of the tapered channel maintains a uniform detection configuration for all cells. The fabricated optofluidic system could detect up to 23 particles per second without becoming clogged; the cell counting efficiency depends on the cell concentration, the flow rate, the ratio of the cell size to the microchannel neck diameter, and the microchannel shape.

When fabricating this kind of optofluidic system, laser direct writing of both microfluidic components and optical waveguides can be performed in the same fabrication step. Therefore, the alignment accuracy of the microchannel and the optical waveguides is limited only by the accuracy of the translation stage. Moreover, there is negligible displacement between the two optical waveguide axes across the microchannel, resulting in high-efficiency detection and manipulation of single cells and microorganisms, as described in this and the following sections.

9.5 Manipulation of Single Cells

Similar optofluidic systems (except that the optical waveguides are parallel to the top surface of the glass chip) have been used for trapping and stretching single cells [22]. To optically trap single cells, two 1070 nm-wavelength light beams with the same energy were coupled into the entrance facets of both waveguides. The distance between the end facets of both waveguides across the microchannel was carefully designed to realize efficient trapping. In addition, microchannels with square cross sections were fabricated by laser direct writing to improve the imaging quality of trapped cells. Trapping and imaging of RBCs were demonstrated, as shown in Fig. 9.5a.

Furthermore, progressive stretching of trapped cells was observed on increasing the optical power of the laser beam, as shown in Fig. 9.5b. The optical stretching technique can be applied to reliably evaluate the mechanical properties of single cells [40].

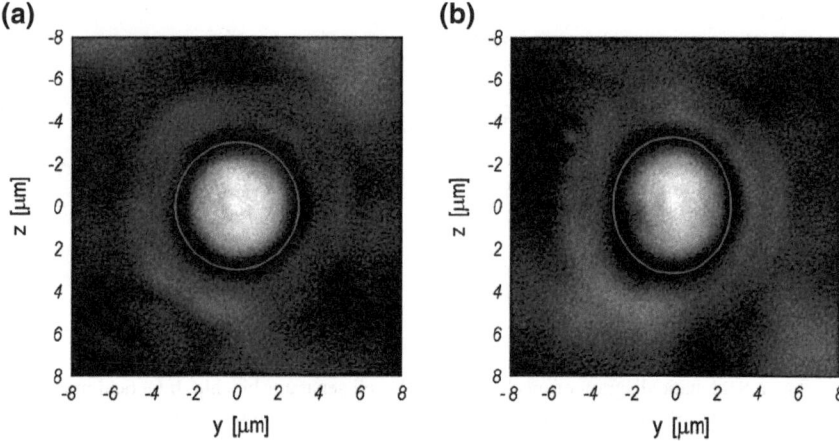

Fig. 9.5 **a** Image of a trapped RBC obtained using a phase contrast microscope. *The red line* represents the recovered contour of the cell. **b** Image of the same RBC elongated by optical forces. RBCs are stretched in the direction parallel to the trapping laser beam [40] (Reproduced with permission from Wiley. ©2007 by Wiley–VCH)

9.6 Cell Sorting

Cell populations often exhibit some kind of heterogeneity, which is problematic for cellular biology experiments. In such experiments, it is important to isolate the species of interest from a heterogeneous population for culturing and genetic analysis. A 3D mammalian cell separator biochip was fabricated by femtosecond laser direct writing of fused silica followed by wet chemical etching [23]. Cell sorting is based on deformability differences resulting from different cell types having different cytoskeletal architectures. Figure 9.6a illustrates the principle of the fabricated biochip for sorting. It consists of a T-junction formed by two microchannels connected by narrow constrictions. These constrictions function as filters for sorting. This structure enables accurate pressure-driven flow control resulting in cell deformation that depends on the cell characteristics. When a heterogeneous population of cells is introduced into this biochip from the inlet, the softer cells are deformed by the pressure gradient maintained across the constrictions and they are guided through the constrictions into outlet 1 of the device. The constriction cross sections should be narrower than the average cell size. The more rigid cells cannot deform sufficiently to pass through the constrictions so that they flow toward outlet 2. A T-junction device with 18 constrictions (see Fig. 9.6b) was employed to demonstrate a cell separator biochip. Human promyelocytic leukemia (HL60) cells were injected into the left microfluidic channel at a constant flow rate of 0.5 ml/min. Healthy looking cells were measured to have an average size of 11.7 ± 1.1 μm with a standard deviation of 1.09 for 25 cells, while the constrictions had a cross section of 4 μm × 8 μm and a length of 200 μm.

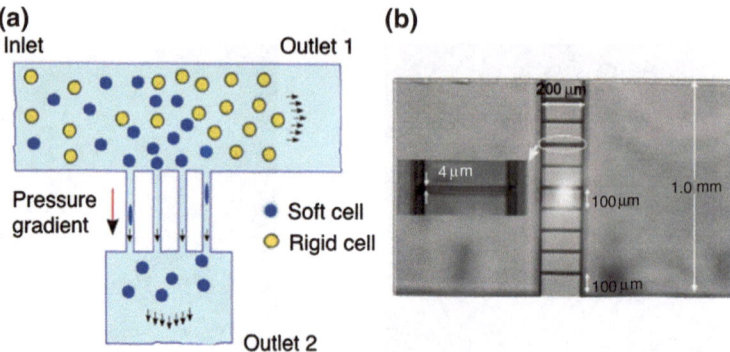

Fig. 9.6 a Schematic diagram of mechanism of 3D cell separator biochip. **b** Optical micrograph of *top view* of constriction array in fabricated biochip [23] (Reproduced with permission from RSC. ©2012 by the Royal Society of Chemistry)

The HL60 cells were successfully collected in the right microfluidic channel passing through the constriction with 81 % of the collected cells viable. This result indicates the possibility of separating a heterogeneous population of cells based on deformability differences between cells.

Another scheme for cell sorting involves utilizing optical forces in combination with fluorescence detection of the cells [25]. Figure 9.7a illustrates the principle of cell sorting using this scheme. Two input channels (INs) merge into a single straight channel where fluorescence detection and sorting are performed. This channel then separates into two output channels (OUTs). The sample liquid containing the cells and buffer solution is introduced from IN1 and IN2, respectively. By adequately controlling the flow rates of the fluids, laminar flow is generated in the single straight channel so that the entire sample with cells is exhausted to OUT1. Applying an optical force pushes cells to the buffer solution side and thus the cells are collected in OUT2. Sorting can be automatically performed based on fluorescence detection of cells. Specifically, the fluorescence laser beam is delivered by the fluorescence waveguide (FWG) to the microchannel, which illuminates the entire height of the microchannel to detect all cells flowing in the microchannel. In this case, the power of the fluorescence laser beam is sufficiently low that it exerts a negligible optical force on the cells. A specific fluorescence signal can be detected when target cells pass through a region in front of the FWG. When a fluorescence signal is detected, the optical force laser beam is automatically switched on. It is guided to the microchannel by the sorting waveguide (SWG), after a moderate delay time to allow for the detected cell to reach the front of the SWG. The target cell is then pushed to the buffer solution side and is sorted into OUT2. A sample consisting of human transformed fibroblasts transfected with a plasmid encoding the enhanced green fluorescent protein (EGFP; excitation wavelength: 488 nm; emission wavelength: 505 nm) was used for the sorting test. About 50 % of these cells were imparted with an intense green fluorescence.

Fig. 9.7 **a** Principle for cell sorting utilizing optical forces combined with fluorescence detection. Demonstration of cell sorting when **b** non-fluorescent and **c** fluorescent cells are detected in an optofluidic device [25] (Reproduced with permission from RSC. ©2012 by the Royal Society of Chemistry)

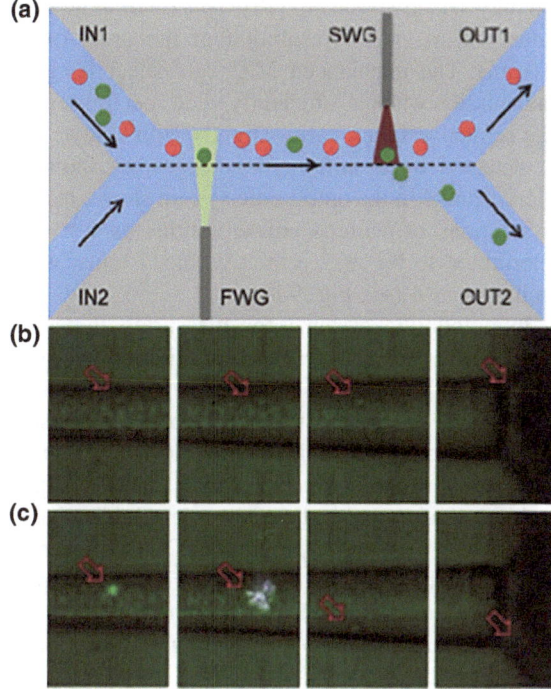

To excite EGFP fluorescence, a 473-nm laser beam was coupled into the FWG. When a non-fluorescent cell was illuminated by this laser beam in the FWG, it did not emit any fluorescence signal and the optical force laser beam remained switched off so that the cell continued to flow and was emitted at OUT1 (Fig. 9.7b). On the other hand, when a fluorescent cell was illuminated, a fluorescence signal was detected. After an appropriate delay time, the optical force laser beam (wavelength: 1070 nm) was switched on and it pushed the cell to the buffer solution side so that the cell was finally sorted to OUT2, as shown in Fig. 9.7(c).

9.7 Concentration Analysis of Liquid Samples

Optofluidic systems in which the microfluidic components are integrated with optical waveguides by femtosecond laser direct writing can be used to measure absorption spectra of liquid samples containing a pH indicator for concentration analysis, as described in Sect. 9.9 [28]. This is a very simple and convenient method for determining the concentrations of various liquid samples.

Another scheme for sensing the concentration of liquid samples by an optofluidic system fabricated by a femtosecond laser employs an unbalanced Mach–Zehnder interferometer (MZI) (see Fig. 8.12 in Chap. 8) to perform label-free, spatially selective sensing [18]. This optofluidic system can measure the refractive index with a spatial resolution of the order of the waveguide mode diameter (11 μm). The unbalanced MZI can detect fringes in the wavelength-dependent transmission when a sufficiently wide spectral region is scanned by a tunable laser. The refractive index varies slightly with analyte concentration; this variation can be detected by a shift in the fringes, as shown in the inset of Fig. 9.8. The microchannel in the optofluidic system shown in Fig. 8.12 was filled with glucose-D solutions of various concentrations as test samples and its sensitivity was determined to be 10^{-4} refractive index units, which corresponds to a detection limit of 4 mM (see Fig. 9.8).

In optofluidic systems for measuring the concentrations of liquid samples introduced above, the sensitivity is limited by the interaction length of the probe beam with the samples, which corresponds to the width of the microfluidic channels. If the probe light could propagate inside the channel, the interaction length would be greatly increased. However, such propagation cannot occur since glass has a higher refractive index than the liquid samples. To overcome this problem, the internal walls of the microfluidic channel were coated with a polymer (Teflon AF, DuPont) by dipping. This polymer has a smaller refractive index (1.31) than that (~ 1.33) of water [29]. Thus, the probe light for liquid analysis can propagate along the inside of the microfluidic channel filled with liquid samples so that the interaction between the liquid and the incident probe light is enhanced, improving the analysis sensitivity. Figure 9.9 shows a 3D illustration of an optofluidic chip fabricated by femtosecond laser direct writing. It consists of a microfluidic channel whose internal walls are coated with a low-refractive-index polymer and an optical waveguide for high sensitivity biochemical liquid assay. The fabricated optofluidic chip was filled with liquid samples. White light from a halogen lamp was coupled by an objective lens to a facet of an optical waveguide

Fig. 9.8 Fringe shift for various glucose-D concentrations (in inset, *solid line* 0 mM; *dashed line* 50 mM; *dotted line* 100 mM) measured by optofluidic system with an unbalanced MZI. Top axis indicates corresponding refractive index change [18] (Reproduced with permission from RSC. ©2010 by the Royal Society of Chemistry)

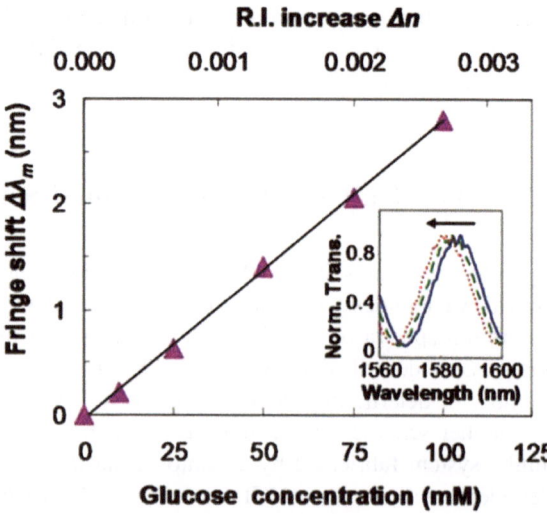

connected to the right side of the microfluidic channel. The light transmitted by the waveguide was introduced to the microfluidic channel filled with a liquid sample and it then propagated inside the channel. Finally, the probe light was reflected at the end of the channel and was directed to an objective lens and a spectrometer. Light propagated inside the liquid-filled channel by total internal reflection due to the coated polymer having a lower refractive index than the liquid sample. To introduce the light into the spectrometer, the sidewall of the left reservoir was tilted at an angle of 10° relative to the optofluidic chip surface. The reflected probe beam was collected by an objective lens and was coupled to the detector head of the external spectrometer for absorbance measurements of the liquid samples in concentration analysis. The sensitivity of the analysis of glucose-D solutions mixed with an enzyme mix as an indicator was significantly improved to a concentration as low as 200 nM, as shown in Fig. 9.10. Furthermore, proteins in bovine serum albumin with concentrations down to 7.5 mM were successfully detected using the Bradford reagent as an indicator.

Most biochips for concentration analysis fabricated by femtosecond laser microprocessing employ light propagating inside the microchannel in the manner described above. However, Maselli et al. have recently demonstrated a refractive index sensor based on evanescent waves [27]. To ensure good overlap between the liquid sample and the evanescent wave, they fabricated a microfluidic channel that was placed in the vicinity of a Bragg grating waveguide (BGW); the distance between the microchannel wall and the waveguide was less than $\sim 2\ \mu m$. Detection of a refractive index change in a liquid sample can be achieved by monitoring the shift in the BGW resonance. This device is capable of resolving refractive index changes of the order of 10^{-4}. Its performance can be enhanced by increasing the quality factor of the Bragg grating cavity or increasing the overlap between the evanescent wave and the liquid sample (e.g., wrapping the BGW with the microfluidic channel wall).

Fig. 9.9 3D illustration of optofluidic chip fabricated by femtosecond laser direct writing. It consists of a microfluidic channel whose internal walls are coated with a low-refractive-index polymer and an optical waveguide for high sensitivity biochemical liquid assay [29]

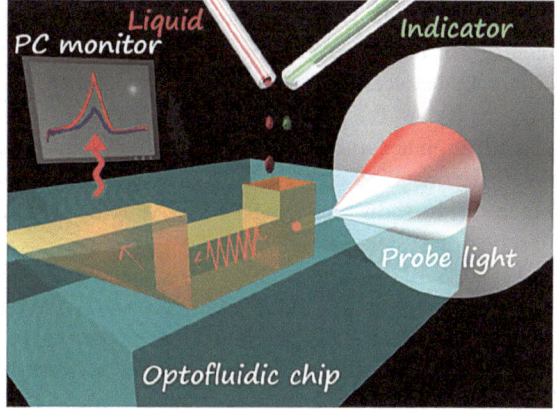

Fig. 9.10 Absorption spectra
of glucose-D solutions with
different concentrations
obtained using the optofluidic
chip shown in Fig. 9.9. An
enzyme mix was added as an
indicator [29]

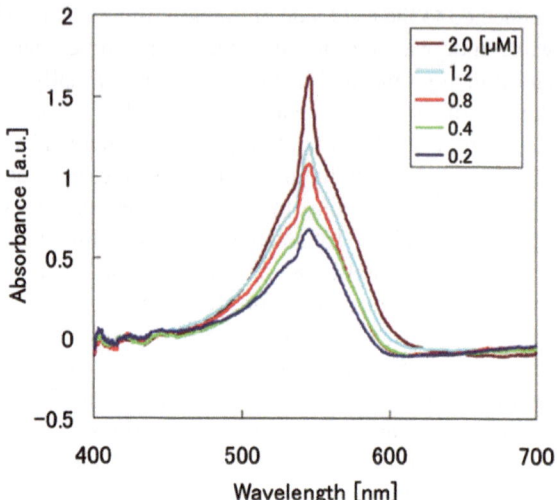

9.8 Rapid Screening of Algae Populations

The detection and identification of submillimeter-sized phytoplankton are of great
importance for monitoring environmental and climate changes, as well as evaluating
water for health reasons. An optofluidic system consisting of a curved waveguide and
a microchannel (see Fig. 9.11a) was fabricated to perform optical classification of
algae species [31–33]. A probe light (1550 nm) delivered by the curved optical
waveguide passed through algae-laden water flow in a microchannel for detection.
The distance between the end of the optical waveguide and the channel is designed to
be sufficiently long that the light emitted from the end facet of the waveguide expands
to illuminate the entire channel height. The curved waveguide increases the signal-
to-noise ratio since it prevents uncoupled light from reaching the photodiode. After
passing through the microchannel, the probe light emitted from the end of the
waveguide was introduced to a four-quadrant detector. The photodetector generated
two signals that depended on the size and shape of the algae species: one is the total
intensity of the detected light ($I_{total} = A + B + C + D$) and the other is the
difference between the two upstream detectors and the two downstream detectors
($\Delta X = (A + B) - (C + D)$) (see Fig. 9.11b for an example). To verify the
classification accuracy of the optofluidic system, a microscope and a camera were
externally installed above the channel and were used to manually identify algae.
Figure 9.11c shows an image of a Cyanothece sample captured by this external
observation system.

Figure 9.12 shows the total (I_{total}) and difference (ΔX) photodiode signals
obtained from nine species of algae, with corresponding micrographs. These
results form the basis of a library for comparing data obtained by the optofluidic
chip. By using a pattern-matching neural network based on the created database,

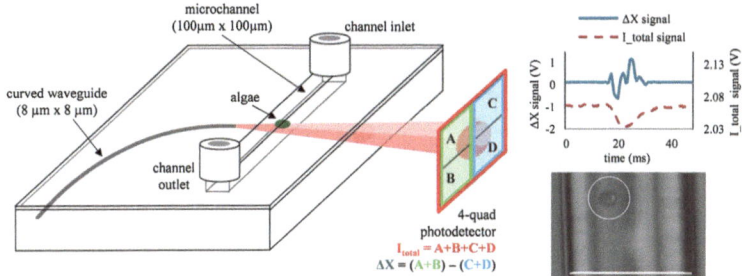

Fig. 9.11 a Schematic of system with a curved waveguide to direct laser light across a microchannel onto a photodetector. **b** Typical obtained signal. **c** Photograph of a Cyanothece sample (*circled*) in microchannel. The channel width (indicated by *white scale bar*) is 100 µm [32] (Reproduced with permission from RSC. ©2012 by the Royal Society of Chemistry)

five algae species were categorized with an average positive identification rate of 78 %. Moreover, the same neural network classification was applied in a field-deployable device for distinguishing the toxin-producing cyanobacterium Cyanothece from detritus in field-collected water; it had a success rate of over 90 %.

9.9 Nanoaquariums for Determining the Functions of Microorganisms

Optofluidic systems have also been used to determine the functions of microorganisms. Such optofluidic systems have been termed nanoaquariums [28, 30]. As one example of nanoaquariums, this section describes the elucidation of the gliding mechanism of *phormidium*. *Phormidium* is a genus of soil-dwelling unicellular, filamentous cyanobacteria, which glide to the seedling roots of vegetables where they form endosymbiotic associations that accelerate vegetable growth. Therefore, it is very important to understand the factors that induce *phormidium* gliding in order to develop methods for accelerating vegetable growth.

To investigate the attraction mechanism that directs *phormidium* to glide toward a seedling, a T-shaped microfluidic channel with three reservoirs at its ends was formed in photosensitive glass. When *phormidium* was introduced into one reservoir and a seedling root into another, the *phormidium* always glided toward the seedling rather than the third empty reservoir. On the other hand, when the third reservoir was filled with carbonic water, the direction to which *phormidium* glided depended on the carbonic water concentration. Additionally, at a critical CO_2 concentration, the cyanobacterium did not glide toward either the seedling root or the carbonic water (below and above this concentration *Phormidium* glided toward the seedling root and the carbonic water, respectively). This suggests that CO_2 secreted during root respiration is a possible attractant. To confirm this hypothesis and to determine the quantity of CO_2 secreted by the seedling roots, an

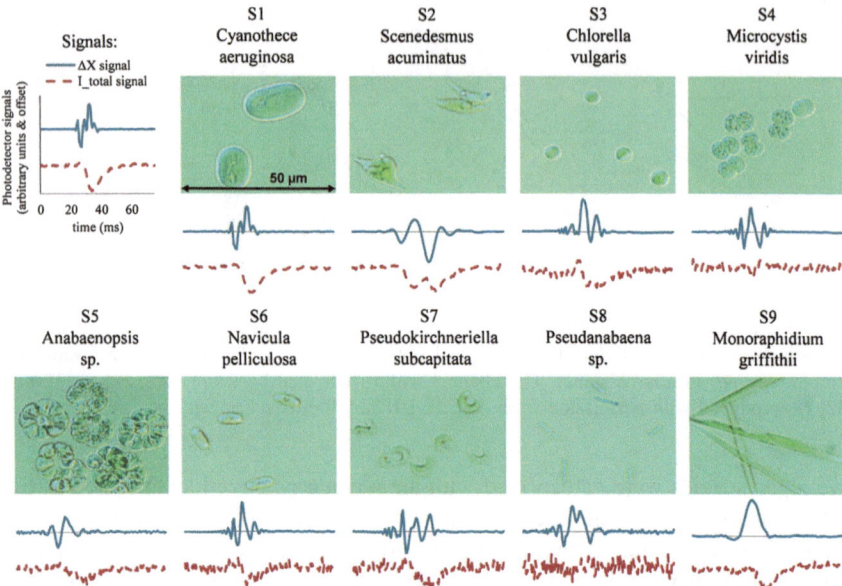

Fig. 9.12 Differential (ΔX) and total (I_{total}) photodiode signals obtained from nine algae species with corresponding micrographs. They form the basis of a library for comparing data obtained by the optofluidic chip [32] (Reproduced with permission from RSC. ©2012 by the Royal Society of Chemistry)

optofluidic system similar to those used to detect and manipulate single cells (as introduced in Sects. 9.4 and 9.5) was fabricated in photosensitive glass. Figure 9.13a shows a schematic diagram of the optofluidic system used in this study. After fabricating a simple straight microfluidic channel in the glass, optical waveguides that intersect the center of the microfluidic channel were written. The microfluidic channel was filled with water containing a pH indicator (bromothymol blue (BTB) solution) and white light from a halogen lamp was coupled to the entrance facet of waveguide I by an objective lens. The white light transmitted by waveguide I passed through the microfluidic channel, which was filled with a liquid sample and was then coupled into optical waveguide II. The light transmitted by waveguide II was coupled into a spectrometer by another objective lens to obtain the absorption spectrum of the sample (Fig. 9.13a). The absorbance was calculated by subtracting the spectrum with the sample in the microfluidic channel from that observed without the sample. The green line in Fig. 9.13b indicates the absorption spectrum of the water containing the BTB solution. It has a large absorption peak at a wavelength of about 620 nm. The intensity of this peak decreases with increasing CO_2 concentration in the water due to the change in the pH. The spectrum of water containing the seedling root (yellow line) is comparable to that of 50 ml water mixed with 15 ml CO_2 (black line). This result implies that the CO_2 concentration generated by root respiration is comparable to that of

Fig. 9.13 Nanoaquarium design and optical absorption measurement for identifying the attractant that induces *Phormidium* gliding. **a** Schematic illustration of microchip integrated with optical waveguides. **b** Optical absorption spectra of water containing BTB solution (*green line*) with a seedling root (*yellow line*) and with carbonic water with different CO_2 concentrations (*blue line* [10 ml CO_2: 50 ml H_2O], *black line* [15 ml CO_2: 50 ml H_2O], and *red line* [25 ml CO_2: 50 ml H_2O]) [28]

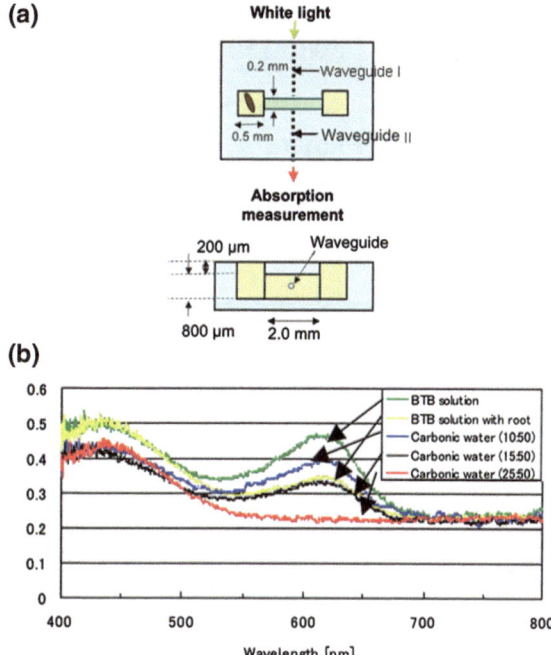

the carbonic water (15 ml CO_2: 50 ml H_2O) used in this experiment. Interestingly, this CO_2 concentration is equal to the critical concentration at which *phormidium* did not glide toward either the seedling root or the carbonic water in the T-shaped microfluidic channel, confirming that CO_2 is the sole attractant for the gliding *phormidlum*.

The gliding movement of *phormidium* has also been found to be sensitive to the illumination conditions (e.g., the intensity and spectral properties of the light). To investigate the effect of illumination, the illuminance in the microchannel was controlled by fabricating optical attenuators around a portion of the microfluidic channel in the nanoaquarium (see Figure 9.14). The optical attenuators were formed by irradiating photosensitive glass with a femtosecond laser beam after fabricating a microfluidic channel. The microchip was then annealed again (but not etched). After annealing, the laser-irradiated regions turned brown due to the formation of lithium metasilicate crystallites, as described in Chap. 6; these regions thus act as optical attenuators. The optical transmission was tuned by varying the number of attenuator layers fabricated. Figure 9.15 shows that *phormidium* glided toward the root in the microfluidic channels when five or fewer optical attenuator layers were stacked, whereas it remained at the channel entrance for more than 2 h when six or more optical attenuator layers were stacked. *Phormidium* glides to a seedling root when white light exceeds a certain intensity; this threshold was estimated to be 1530 lx based on the illuminance of the optical

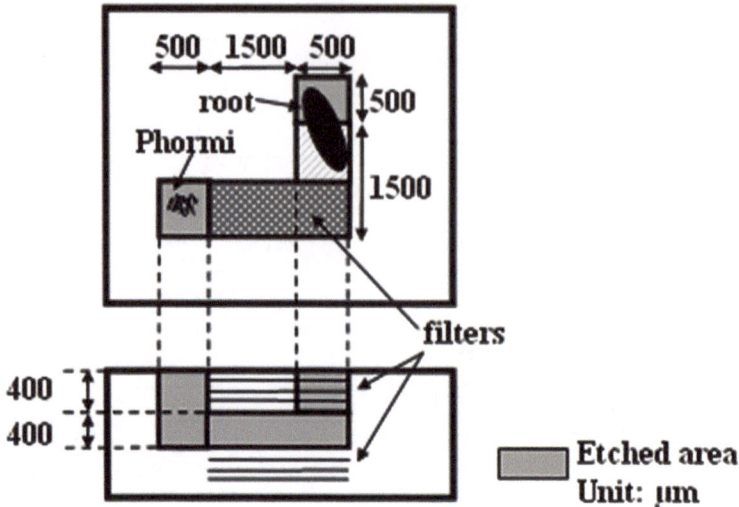

Fig. 9.14 Structural diagrams of nanoaquarium integrated with optical attenuators. The same number of optical attenuator layers were formed above and below the region of the microfluidic channel to control the white light intensity in the microfluidic channel [28]

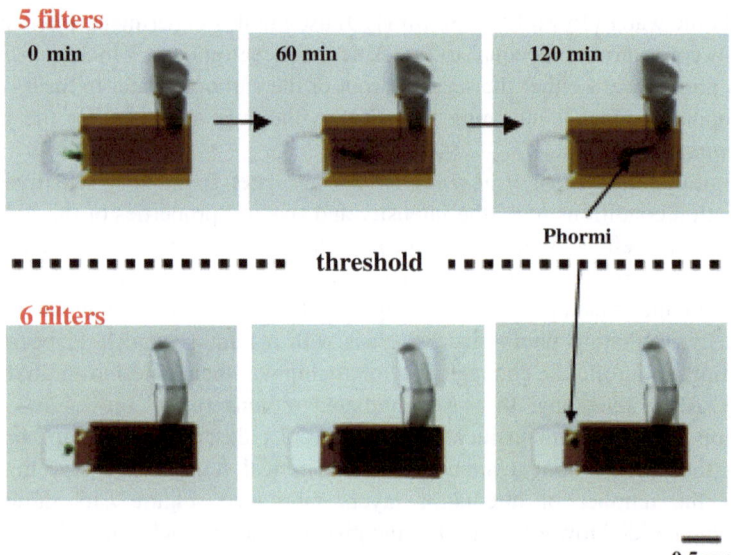

Fig. 9.15 Sequential microscope images of gliding movement in nanoaquarium integrated with optical filters. *Phormidium* glided to the seedling root in the microfluidic channel covered with up to five optical attenuator layers, whereas it did not glide to the root when there were six optical attenuator layers, even after 2 h [28]

microscope and the transmittance of six optical attenuator layers. Further investigation using band-pass filters revealed that only light in the red wavelength region (i.e., in the spectral range 640–700 nm) promotes *phormidium* gliding. These results are important for developing methods for accelerating vegetable seedling growth.

9.10 Summary

Femtosecond lasers are powerful tools for fabricating truly 3D microfluidic systems with complex structures inside glass without using stacking and bonding processes. These microfluidic systems can perform highly efficient fluid mixing and cell sorting. Furthermore, femtosecond lasers can be used to fabricate other functional microcomponents (e.g., micro-optical components, microelectronic components, and micromechanics) and integrate them in a single glass chip. The microcomponents can be easily aligned. Thus, biochips that are highly functionalized for high efficiency and high sensitivity biochemical experiments can be produced. Metallization of glass enabled highly sensitive sensing of molecules based on SERS. Integration of micro-optical components including optical waveguides and optical filters has been utilized in many biological applications such as single cell detection, manipulation and sorting of cells, concentration analysis of liquid samples, and detection and elucidation of functions of microorganisms and bacteria.

Despite the great effort that has been expended in investigating applications of biochips fabricated by femtosecond lasers, femtosecond laser processing is still not as popular as conventional techniques for fabricating biochips. Nevertheless, femtosecond laser processing is promising due to its ability to realize one-step integration of microfluidics and other microcomponents in glass chips. Glass is an ideal substrate material for many applications mainly because of its high chemical inertness and high optical transparency. The ultimate dream of femtosecond laser integration of biochips is to create a complete "all-in-one" lab-on-a-chip or optofluidic system in which all the functional components are three-dimensionally integrated in a single exposure step by femtosecond laser direct writing followed by optional batch post-processing procedures, which are cost-effective for mass-production. This is certainly a formidable challenge, but significant progress has been made toward realizing this goal and important technical advances have been demonstrated, as described in this chapter. Another problem is that femtosecond laser processing has long been regarded as an expensive and time-consuming technique; this image hampers its widespread use. Fortunately, the rapid development of next-generation femtosecond lasers in recent years that have high stability, high reliability, high power, and ease of operation at lower cost will promote the use of femtosecond lasers for fabricating biochips.

References

1. Marcinkevičius A, Juodkazis S, Watanabe M et al (2001) Femtosecond laser-assisted three-dimensional microfabrication in silica. Opt Lett 26:277–279
2. Masuda M, Sugioka K, Cheng Y et al (2003) 3-D microstructuring inside photosensitive glass by femtosecond laser excitation. Appl Phys A 76:857–860
3. Li Y, Itoh K, Watanabe W et al (2001) Three-dimensional hole drilling of silica glass from the rear surface with femtosecond laser pulses. Opt Lett 26:1912–1914
4. Liao Y, Ju Y, Zhang L et al (2010) Three-dimensional microfluidic channel with arbitrary length and configuration fabricated inside glass by femtosecond laser direct writing. Opt Lett 35:3225–3227
5. Davis KM, Miura K, Sugimoto N et al (1996) Writing waveguides in glass with a femtosecond laser. Opt Lett 21:1729–1731
6. Cheng Y, Sugioka K, Midorikawa K et al (2003) Three-dimensional micro-optical components embedded in photosensitive glass by a femtosecond laser. Opt Lett 28:1144–1146
7. Wang Z, Sugioka K, Midorikawa K (2007) Three-dimensional integration of microoptical components buried inside photosensitive glass by femtosecond laser direct writing. Appl Phys A 89:951–955
8. Sugioka K, Hongo T, Takai H et al (2005) Selective metallization of internal walls of hollow structures inside glass using femtosecond laser. Appl Phys Lett 86:171910
9. Hanada Y, Sugioka K, Midorikawa K (2008) Selective metallization of photostructurable glass by femtosecond laser direct writing for biochip application. Appl Phys A 90:603–607
10. Xu J, Liao Y, Zeng HD et al (2007) Selective metallization on insulator surfaces with femtosecond laser pulses. Opt Express 15:12743–12748
11. Masuda M, Sugioka K, Cheng Y et al (2004) Direct fabrication of freely movable microplate inside photosensitive glass by femtosecond laser for lab-on-chip application. Appl Phys A 78:1029–1032
12. Kiyama S, Tomita T, Matsuo S et al (2009) Laser fabrication and manipulation of an optical rotator embedded inside a transparent solid material. J Laser Micro/Nanoeng 4:18–21
13. Sugioka K, Cheng Y (2012) Femtosecond laser processing for optofluidic fabrication. Lab Chip 12:3576–3589
14. Osellame R, Hoekstra HJWM, Cerullo1 G et al (2011) Femtosecond laser microstructuring: an enabling tool for optofluidic lab-on-chips. Laser Photonics Rev 5:442–463
15. Wang Z, Sugioka K, Midorikawa K (2008) Fabrication of integrated microchip for optical sensing by femtosecond laser direct writing of foturan glass. Appl Phys A 93:225–229
16. Liao Y, Xu J, Cheng Y et al (2008) Electro-optic integration of embedded electrodes andwaveguides in LiNbO3 using a femtosecond laser. Opt Lett 33:2281–2283
17. Cheng Y, Sugioka K, Midorikawa K et al (2004) Microfluidic laser embedded in glass by three-dimensional femtosecond laser microprocessing. Opt Lett 29:2007–2009
18. Crespi A, Gu Y, Ngamsom B et al (2010) Three-dimensional Mach-Zehnder interferometer in a microfluidic chip for spatially-resolved label-free detection. Lab Chip 10:1167–1173
19. Zhou Z, Xu J, He F et al (2010) Surface-enhanced Raman scattering substrate fabricated by femtosecond laser induced co-deposition of silver nanoparticles and fluorescent molecules. Jpn J Appl Phys 49:022703
20. Liao Y, Song J, Li E et al (2012) Rapid prototyping of three-dimensional microfluidic mixers in glass by femtosecond laser direct writing. Lab Chip 12:746–749
21. Kim M, Hwang DJ, Jeon H et al (2009) Single cell detection using a glass-based optofluidic device fabricated by femtosecond laser pulses. Lab Chip 9:311–318
22. Bellini N, Vishnubhatla KC, Bragheri F et al (2010) Femtosecond laser fabricated monolithic chip for optical trapping and stretching of single cells. Opt Express 18:4679–4688
23. Choudhury D, Ramsay WT, Kiss R et al (2012) A 3D mammalian cell separator biochip. Lab Chip 12:948–953

24. Applegate RW Jr, Squier J, Vestad T et al (2006) Microfluidic sorting system based on optical waveguide integration and diode laser bar trapping. Lab Chip 6:422–426
25. Brahheri F, Minzioni P, Vazquez RM et al (2012) Optofluidic integrated cell sorter fabricated by femtosecond lasers. Lab Chip 12:3779–3784
26. Bragheri F, Ferrara L, Bellini N et al (2010) Optofluidic chip for single cell trapping and stretching fabricated by a femtosecond laser. J Biophotonics 3:234–243
27. Maselli V, Grenier JR, Ho S et al (2009) Femtosecond laser written optofluidic sensor: Bragg grating waveguide evanescent probing of microfluidic channel. Opt Express 17:11719–11729
28. Hanada Y, Sugioka K, S-Ishikawa et al (2011) 3D microfluidic chips with integrated functional microelements fabricated by a femtosecond laser for studying the gliding mechanism of cyanobacteria. Lab Chip 11:2109–2115
29. Hanada Y, Sugioka K, Midorikawa K (2012) Highly sensitive optofluidic chips for biochemical liquid assay fabricated by 3D femtosecond laser micromachining followed by polymer coating. Lab Chip 12:3639–3688
30. Hanada Y, Sugioka K, Kawano H et al (2008) Nano-aquarium for dynamic observation of living cells fabricated by femtosecond laser direct writing of photostructurable glass. Biomed Microdevices 10:403–410
31. Schaap A, Bellouard Y, Rohrlack T (2011) Biomed. Optofluidic lab-on-a-chip for rapid algae population screening. Opt Express 2:658–664
32. Schaap A, Rohrlack T, Bellouard Y (2012) Optical classification of algae species with a glass lab-on-a-chip. Lab Chip 12:1527–1532
33. Schaap A, Rohrlack T, Bellouard Y (2012) Lab on a chip technologies for algae detection: a review. J Biophotonics 5:8–9
34. Lan X, Han Y, Wei T et al (2009) Surface-enhanced Raman-scattering fiber probe fabricated by femtosecond laser. Opt Lett 34:2285–2287
35. Han Y, Lan X, Wei T et al (2009) Surface enhanced Raman scattering silica substrate fast fabrication by femtosecond laser pulses. Appl Phys A 97:721–724
36. http://www.kirbyresearch.com/index.cfm/page/ri/ufluids.htm
37. Wiggins S, Ottin JM (2004) Foundations of chaotic mixing. Philos Trans R Soc A 362:937–970
38. Carrière P (2007) On a three-dimensional implementation of the baker's transformation. Phys Fluids 19:118110
39. Ju Y, Liao Y, Zhang L et al (2012) Fabrication of large-volume microfluidic chamber embedded in glass using three-dimensional femtosecond laser micromachining. Microfluid Nanofluid 11:111–117
40. Lincoln B, Schinkinger S, Travis K et al (2007) Reconfigurable microfluidic integration of a dual-beam laser trap with biomedical applications. Biomed Microdevices 9:703–710

Chapter 10
Summary and Outlook

Abstract The primary goal of this book is to comprehensively review state-of-the-art femtosecond laser three-dimensional (3D) micromachining techniques for microfluidic and optofluidic applications, including techniques for fabricating microfluidic components, optical waveguides, free-space micro-optical components, microelectrodes, and integrated optofluidic systems and devices. It also presents typical examples of applications of femtosecond-laser-fabricated microfluidic and optofluidic chips for chemical sensing and investigating biological species. Comparison with conventional lithography-based fabrication techniques reveals the uniqueness and versatility of femtosecond laser micromachining. In this chapter, we summarize the results and contributions presented in this book and overview the future outlook of this field.

10.1 Introduction

Integrated microfluidic and optofluidic systems have found important applications ranging from chemical and biological sensing to reconfigurable photonics. Today, most microfluidic chips are manufactured using lithography-based two-dimensional (2D) microfabrication techniques such as soft lithography using poly(dimethylsiloxane) (PDMS) substrates, which lack the ability to directly form 3D microfluidic networks in transparent substrates without stacking and bonding. Femtosecond laser direct writing is presently the only technique used to modify the interiors of transparent materials and it thus opens up new avenues for fabricating a variety of microfluidic and optofluidic devices. This final chapter summarizes the previous chapters and reviews the future prospects of this young, yet rapidly evolving research area.

K. Sugioka and Y. Cheng, *Femtosecond Laser 3D Micromachining for Microfluidic and Optofluidic Applications*, SpringerBriefs in Applied Sciences and Technology, DOI: 10.1007/978-1-4471-5541-6_10, © The Author(s) 2014

10.2 Summary

Chapter 1 briefly described the background of microfluidic and optofluidic systems and introduced the subsequent chapters. Chapter 2 provided an overview of different approaches that have been developed for fabricating microfluidic and optofluidic components and devices on various substrate materials (e.g., polymers, glass, silicon, and paper). The advantages and disadvantages of these approaches were outlined and compared. Chapter 3 reviewed the fundamentals and characteristics of femtosecond laser processing. It also introduced state-of-the-art femtosecond laser processing. Chapters 4, 5, 6 and 7, described the technical details for using femtosecond laser direct writing to fabricate microfluidic, micromechanic, micro-optic, and microelectronic components in glass, respectively. Examples of fabricated structures were described, including microfluidic channels and chambers (Chap. 4), a microvalve and a micropump (Chap. 5), optical waveguides, micromirrors, microlenses, and optical attenuators (Chap. 6), and microelectrodes and a micro-heater (Chap. 7). These functional components can be monolithically integrated in a single glass substrate, as described in Chap. 8. Chapter 9 described examples of applications of microfluidic and optofluidic systems fabricated by femtosecond laser processing, such as surface-enhanced Raman scattering spectroscopy, a micromixer, single cell detection, concentration analysis of liquid samples, and nanoaquariums.

10.3 Outlook

Past efforts in this research field have mainly been in two directions: development of femtosecond laser fabrication techniques and using fabricated systems in a wider range of applications. The former has resulted in the performance and functionality of microfluidic systems and integrated optofluidic devices being substantially enhanced, whereas the latter has led to novel devices and applications in the fields of life science, chemistry, and photonic technology. This general trend will continue in the future, although the latter direction may be emphasized more.

On the technical development side, although femtosecond laser direct writing has been demonstrated to be a powerful tool for fabricating microfluidic and optofluidic systems, there is still room for improvement. First, the current fabrication resolution for internal 3D modification in glass is of the order of the wavelength of light, mainly due to the diffraction limit of the focusing system, heat diffusion in the laser affected zone, and resolution degradation that occurs in post-exposure fabrication processes such as wet chemical etching. This resolution is too low for cutting edge fluidic and photonic applications such as nanofluidics and nanophotonics [1, 2]. To overcome this long-standing problem, it is essential to better understand the microscopic mechanisms of ultrafast energy deposition and transport dynamics in microscale focal volumes, which can be extremely sensitive to the spatiotemporal structure of the focused femtosecond laser pulses [3–5].

When this has been accomplished, subwavelength resolutions that exceed the diffraction limit are likely to be realized using sophisticated spatiotemporal pulse shaping or focal spot engineering techniques that permit precise control of the dynamics of energy deposition in glass within the focal volume with a sub-wavelength spatial resolution [6–8]. Second, with either femtosecond-laser-assisted wet chemical etching or water-assisted femtosecond laser 3D drilling, the length of the microfluidic structures directly fabricated in glass is limited to a few millimeters to ~ 1 cm [9]. In contrast, for 2D lithography techniques such as soft lithography, there is in principle no limit on the length of microfluidic chips [10]. Fortunately, this bottleneck has recently been resolved by direct writing of 3D microfluidic channels with nearly arbitrary lengths and configurations (i.e., channels with total lengths up to ~ 10 cm have been demonstrated) in porous glass using focused femtosecond laser pulses [11]. A 3D microfluidic mixer with a high mixing efficiency has been demonstrated using this technique, while other applications have yet to be explored [12]. Finally, femtosecond laser processing has long been regarded as an expensive and time-consuming technique; this image has hampered the widespread use of this technique. However, this situation is currently undergoing a dramatic change due to the rapid development of next-generation femtosecond lasers that offer higher fabrication efficiency and operating stability at reduced costs [13]. In the future, ultrafast laser materials processing will benefit from the rapid development of ultrafast laser systems and ultrafast laser techniques.

On the application side, as a field that is still in its infancy, femtosecond laser processing is currently not as popular as conventional techniques for optofluidic applications, despite its unparalleled capabilities for 3D fabrication and integration. In addition, most optofluidic devices produced by femtosecond laser processing require straightforward incorporation of optical functions in microfluidic systems to enhance their sensing capabilities. Many new directions remain to be explored. For example, fabrication of tunable optical systems by synergistically combining fluidic and optical components has been intensively investigated since the birth of optofluidics [14, 15]. Using femtosecond laser processing, such integrated devices can be directly fabricated in glass without post-assembling. The use of glass substrates has the potential to realize optical performance and chemical stability that are superior to those of polymers. Another fascinating opportunity is provided by the emerging optofluidic technique for sunlight-based energy applications where optical waveguides can be incorporated into photobioreactors or photocatalytic systems to improve either the sunlight collection or distribution performance [16]. Femtosecond laser processing is attractive due to its ability to realize one-step integration of microfluidic and micro-optical components in glass. The selective metallization technique described in Chaps. 7–10 can also be used to produce optofluidic systems with high sensing performance based on surface-enhanced Raman scattering [17]. Glass is an ideal substrate material for such applications mainly because of its high chemical inertness. The ultimate goal of femtosecond laser integration of optofluidics is to create a complete "all-in-one" lab-on-a-chip or optofluidic system in which all the functional components are simultaneously integrated in a single exposure step by femtosecond laser direct writing, followed by optional batch post-processing procedures, which are cost-effective for mass-production [18, 19]. This is certainly a formidable challenge, but significant progress has been made toward realizing this goal and important technical advances have been realized [20–23].

The future development of this research field will benefit the optofluidic and biochemistry communities. With future development and refinement, microfluidic fabrication and optofluidic integration by femtosecond laser processing will not only be suitable for constructing tailor-made research devices but also for realizing mass production.

References

1. Napoli M, Eijkel JCT, Pennathur S (2010) Nanofluidic technology for biomolecule applications: a critical review. Lab Chip 10:957–985
2. Abgrall P, Nguyen NT (2008) Nanofluidic devices and their applications. Anal Chem 80:2326–2341
3. Kazansky PG, Yang W, Bricchi E et al (2007) "Quill" writing with ultrashort light pulses in transparent materials. Appl Phys Lett 90:151120(3)
4. Vitek DN, Block E, Bellouard Y (2010) Spatio-temporally focused femtosecond laser pulses for nonreciprocal writing in optically transparent materials. Opt Express 18:24673–24678
5. Yang WJ, Kazansky PG, Svirko YP (2008) Non-reciprocal ultrafast laser writing. Nat Photonics 2:99–104
6. He F, Xu H, Cheng Y et al (2010) Fabrication of microfluidic channels with a circular cross section using spatiotemporally focused femtosecond laser pulses. Opt Lett 35:1106–1108
7. Durfee CG, Greco M, Block E et al (2012) Intuitive analysis of space-time focusing with double-ABCD calculation. Opt Express 20:14244–14259
8. Stoian R, Boyle M, Thoss A et al (2002) Laser ablation of dielectrics with temporally shaped femtosecond pulses. Appl Phys Lett 80:353–355
9. Kiyama S, Matsuo S, Hashimoto S et al (2009) Examination of Etching Agent and Etching Mechanism on Femotosecond Laser Microfabrication of Channels Inside Vitreous Silica Substrates. J Phys Chem C 113:11560–11566
10. Thorsen T, Maerkl SJ, Quake SR (2002) Microfluidic Large Scale Integration. Science 298:580–584
11. Liao Y, Ju Y, Zhang L et al (2010) Three-dimensional microfluidic channel with arbitrary length and configuration fabricated inside glass by femtosecond laser direct writing. Opt Lett 35:3225–3227
12. Liao Y, Song J, Li E (2012) Rapid prototyping of three-dimensional microfluidic mixers in glass by femtosecond laser direct writing. Lab Chip 12:746–749
13. Tünnermann A, Schreiber T, Limpert J et al (2010) Fiber lasers and amplifiers: an ultrafast performance evolution. Appl Opt 49:F71–F78
14. Psaltis D, Quake SR, Yang C (2006) Developing optofluidic technology through the fusion of microfluidics and optics. Nature 442:381–386
15. Monat C, Domachuk P, Eggleton BJ (2007) Integrated optofluidics: a new river of light. Nat Photonics 1:106–114
16. Erickson D, Sinton D, Psaltis D (2011) Optofluidics for energy applications. Nat Photonics 5:583–590
17. Zhou Z, Xu J, Cheng Y (2008) Surface-enhanced Raman scattering substrate fabricated by femtosecond laser direct writing. Jpn J Appl Phys 47:189–192

18. Sugioka K, Cheng Y, Midorikawa K (2007) "All-in-One" Chip Fabrication by 3D Femtosecond Laser Microprocessing for Biophotonics. J Phys: Conf Ser 59:533–538
19. Balslev S, Jorgensen AM, Bilenberg B (2006) Lab-on-a-chip with integrated optical transducers. Lab Chip 6:213–217
20. Sugioka K, Cheng Y (2011) Integrated microchips for biological analysis fabricated by femtosecond laser direct writing. MRS Bull 36:1020–1027
21. Osellame R, Hoekstra HJWM, Cerullo1 G et al (2011) Femtosecond laser microstructuring: an enabling tool for optofluidic lab-on-chips. Laser Photonics Rev 5:442–463
22. Schaap A, Rohrlack T, Bellouard Y (2012) Lab on a chip technologies for algae detection: a review. J Biophotonics 5:8–9
23. Sugioka K, Cheng Y (2012) Femtosecond laser processing for optofluidic fabrication. Lab Chip 12:3576–3589